PRACTICAL APPLICATION OF AZOLLA FOR RICE PRODUCTION

Developments in Plant and Soil Sciences

Volume 13

Practical Application of Azolla for Rice Production

Proceedings of an International Workshop, Mayaguez, Puerto Rico, November 17–19, 1982

Edited by

W.S. SILVER
Department of Biology
University of South Florida
USA

E.C. SCHRÖDER
Department of Agronomy
University of Puerto Rico

1984 **MARTINUS NIJHOFF/DR W. JUNK PUBLISHERS**
a member of the KLUWER ACADEMIC PUBLISHERS GROUP
DORDRECHT / BOSTON / LANCASTER

Distributors

for the United States and Canada: Kluwer Academic Publishers, 190 Old Derby Street, Hingham, MA 02043, USA
for the UK and Ireland: Kluwer Academic Publishers, MTP Press Limited, Falcon House, Queen Square, Lancaster LA1 1RN, England
for all other countries: Kluwer Academic Publishers Group, Distribution Center, P.O. Box 322, 3300 AH Dordrecht, The Netherlands

Library of Congress Cataloging in Publication Data

Main entry under title:

Practical application of azolla for rice production.

(Developments in plant and soil sciences ; 13)
First International Workshop on "Practical Applica-
tions of Azolla for Rice Production", organized by the
University of Puerto Rico-Mayaguez Campus.
Includes bibliographies and index.
1. Rice--Fertilizers--Congresses. 2. Azolla as
fertilizer--Congresses. I. Silver, W. S. II. Schröder,
E. C. III. International Workshop on "Practical
Applications of Azolla for Rice Production" (1st :
1982 : Mayagüez, P.R.) IV. University of Puerto

Rico (Mayagüez Campus) V. Seri s.
S667.R5P73 1984   633.1'88947   84-16585
ISBN-13:978-94-009-6203-3   e-ISBN-13:978-94-009-6201-9
DOI: 10.1007/978-94-009-6201-9

Copyright

PREFACE

The production of rice has increased considerably in recent years due to the release of improved varieties and the adoption of better fertilization practices. Nevertheless, the production and use of inorganic N fertilizer involves costly investments in terms of energy and transport, the need for complex manufacturing plants, as well as the potential for environmental pollution.

The use of agricultural systems that include dinitrogen fixing organisms appears to be an economically sound cultural practice. In the particular case of rice, biological nitrogen fixation by Azolla, blue-green algae (BGA), and heterotrophic microorganisms has long been recognized, in southeast Asia, as a fertilizer for rice culture. The Azolla-Anabaena association has the unique property of being able to retain a significant amount of nitrogenase activity in the presence of combined nitrogen, making the system compatible with inorganic nitrogen fertilization.

Researchers working with Azolla (N_2 fixation) are dispersed in countries of southeast Asia, Africa, Europe and Latin America, making it difficult to share ideas, concepts and research results on a more personal basis. Considering the potential positive impact of growing rice in association with Azolla, and the lack, to date, of an international gathering of scientists dedicated to Azolla research, the First International Workshop on "Practical Applications of Azolla for Rice Production" was organized by the University of Puerto Rico-Mayaguez Campus.

The workshop was held November 17-19, 1982, at the Mayaguez Campus of the University of Puerto Rico. Two days of invited contributions, paper sessions and round tables were followed by a field trip to the rice growing area and mill located at Arecibo, Puerto Rico.

This volume contains most of the papers presented at the meeting and others submitted later, as well as those abstracts where full papers were not available.

The U.S. Agency of International Development provided the financial support through its BNF Project and Title XII Strengthening Grant Program. At the local level, the Government of Puerto Rico through its Rice Development Corporation and private companies co-sponsored the workshop.

Tampa and Mayaguez, July 1984

W. S. Silver and E. C. Schröder

Contents

VIII

1. Azolla-Anabaena Symbioses: Basic Biology, Use, and Prospects
   for the Future

G. A. Peters
C.F. Kettering Research Laboratory
150 E. South College Street
Yellow Springs, Ohio 45387, USA

Key words   Anabaena   Azolla   Heterocyst   Photosynthesis   Sporophytes
Symbiosis

1.   Introduction

For the benefit of newcomers to the area of Azolla research some
introductory material is presented.   Azolla is a genus of free-
floating, heterosporous aquatic pteridophytes.   At present the
genus usually is considered to contain six extant species in two
sections.   The subgenus Euazolla includes A. filiculoides Lamarck,
A. caroliniana Willdenow, A. mexicana Presl and A. microphylla
Kaulfuss, while the subgenus Rhizosperma includes A. pinnata R.
Brown and A. nilotica DeCaisne.   Species identification, which
is in need of further assessment and possible revision, is based
primarily upon features of the reproductive structures.   The four
new world species in the Euazolla are characterized by three megaspore
floats while the two old world species in the Rhizosperma have
nine megaspore floats.
   All   of   the   Azolla   species   normally   contain   an N_2-fixing
cyanobacterium as an endophyte.   The endophyte occupies specialized
cavities formed in the dorsal leaf lobes of the fern and can provide
the associations with their total N requirement by fixation of
atmospheric N_2.   The endophytic cyanobacterium belongs to the
Nostocaceae and is commonly referred to as Anabaena azollae
Strasburger.   The actual degree of relatedness of the endophyte
in the individual Azolla species is speculative and constitutes
an area of current research.
   The Azolla species are widely distributed in tropical and temperate
fresh water environments.   They do not colonize swiftly moving or
large open bodies of water since wind and wave action as well as
other forms of turbulence cause fragmentation and diminished growth.
Instead, they are found on the more placid surfaces of ponds,
canals, marshes and rice paddies.   If other nutritional factors
are adequate, Azolla-Anabaena associations can exhibit rapid
vegetative growth in areas where combined N levels are limiting
to other aquatic plants, subsequently providing a source of combined
N to the environment.   This attribute has been recognized for many
years in Vietnam and China where Azolla traditionally has been
grown as a biofertilizer or green manure crop for lowland rice.
   My personal involvement with basic research on these symbiotic
associations has spanned the past decade.   During this period
there has been a marked increase in consideration of their potential
to offset the fertilizer N requirements of rice in many areas.
   This was predicated largely by the energy crisis of the mid-1970's
and Azolla subsequently has been undergoing a metamorphosis from

an undersirable water weed to a potentially useful crop in many areas where lowland rice is grown. In fact, the increased lay awareness, scientific interest, and research conducted on these associations during the past five years has been remarkable.

As with any agricultural crop, the effective use of Azolla necessitates an understanding of their biology with a view toward improvement. Integration of useful new information from basic and applied research, appropriate management procedures, and common sense all will play a role in circumventing or minimizing current limitations to the realization of their full potential.

I would emphasize two points. First, this is an overview and is not intended to constitute a comprehensive review of the Azolla literature. (For comprehensive reviews see 2, 19, 20, 28, 29, 37, and inclusive citations). Second, the following assessment of the basic biology, current use and future prospects of these associations will be restricted to A. caroliniana, A. filiculoides, A. mexicana and A. pinnata. I have had limited personal experience with A. nilotica and A. microphylla and to date I am unaware of any published field work using either of them.

2. Basic Biology

2.1 Sporophytes

The sporophytes consist of prostrate, multibranched floating stems which bear deeply bilobed leaves and adventitious roots. The ventral leaf lobes are only one cell thick at their distal end are nearly achlorophyllous in the absence of a combined N source, and float on the water surface. The dorsal lobes are aerial.

At maturity they exhibit a chlorophyllous, multilayered mesophyll bordered by epidermal cells and a well defined cavity which is located in the proximal half. The cavity is characterized by two types of epidermal trichomes, is occupied by the endophytic Anabaena, and is formed by an infolding of the adaxial epidermis.

While it is within the leaf, the endophyte within the cavity is extracellular. The abaxial epidermis of the dorsal lobes has numerous stomata and bears single-celled papillae. These papillae may be a species specific characteristic and merit further assessment under a variety of growth conditions. The development of the adventitious roots is extremely well ordered and has been described in considerable detail for A. pinnata (8).

2.2 Morphology of the Symbiosis

The synchronous development of the Anabaena with that of the Azolla leaves and the developmental profile occurring along a main stem axis was initially reported by Hill (11). The sporophytes have a dorsi-ventral organization. Each stem apex curves upward, away from the water surface, and has a small colony of undifferentiated Anabaena filaments associated with it. The stem apices and associated Anabaena cells are protected by an overarching of the differentiated dorsal and ventral leaf lobes and are completely isolated from the external environment. Growth of the apical Anabaena colonies appears to be entirely coordinated with that of the Azolla.

Establishment of Anabaena filaments in the cavity of each leaf begins in the developing leaves contiguous with the apical Anabaena colony. The partitioning of the Anabaena filaments appears to involve a specialized epidermal trichome termed the primary branched hair (PBH) (4, 29). This hair originates from the axil of a leaf primordium on the apical meristem and rapidly differentiates a single stalk cell and several elongate terminal cells which grow toward the Anabaena colony. Filaments of the endophyte become entwined around the terminal cells and with subsequent leaf development the PBH and entwined Anabaena filaments become positioned at the proximal end of a depression representing the initial stage in the formation of the leaf cavity. Cells around the rim of the depression then produce epidermal cells which close the cavity, engulfing the primary branched hair and Anabaena filaments. This sequence of events is effectively complete by the time the youngest dorsal leaf lobe is discernible to the naked eye (4, 29).

As a leaf cavity is closing, the endophyte begins to differentiate heterocysts and additional epidermal hairs are formed on the cavity wall. One of these hairs is quite similar to the PBH. However, it differentiates fewer terminal cells, is located at the back of the cavity depression, and has been termed the secondary branched hair, (SBH) (4). The two branched hairs are located in similar positions in every leaf cavity, always on the path of the foliar trace. All the other hairs are comprised of only two cells: a stalk cell and a terminal cell. They constitute a population distinct from the branched hairs and are termed simple hairs (4). A mature cavity in A. caroliniana may contain up to 25 simple hairs, randomly distributed around the cavity wall except for the lower distal quadrant.

The observations (7, 41) that epidermal hairs of the leaf cavities exhibit transfer cell ultrastructure (24) have been verified for both hair populations (4). Additional ultrastructural studies (29, Calvert et al., in preparation) have revealed a developmental aspect of the ultrastructure in both populations of epidermal hairs. Cell wall elaborations associated with transfer cell ultrastructure can be distinguished also in the two or three cell layers which separate the site of branched hair attachment on the cavity wall from the foliar trace; and transfer cell ultrastructure is well developed within the foliar trace, especially in the xylem parenchyma, in the region of each branched hair (Calvert et al., in preparation).

In general, our observations of the endophyte are consistent with the observations of Hill (12). Filaments of the apical Anabaena colonies lack heterocysts. As the filaments become associated with the PBH and engulfed in the forming leaf cavities, heterocysts are differentiated. During leaf maturation cell division in the endophyte lessens, the individual cells enlarge, and differentiation of heterocysts continues. In A. caroliniana akinetes are also differentiated. In mature cavities the heterocyst frequency is about 30%, the vegetative cells of the Anabaena are filled with photosynthetic lamellae, and there are two branched hairs and about 25 simple hairs. Anabaena filaments occupy the entire area of developing cavities but in mature cavities the filaments are located around the periphery of those portions of the cavity wall bearing epidermal hairs and adjacent to photosynthetic mesophyll.

3. Physiological Aspects

3.1 Photosynthesis

The eukaryotic Azolla and the prokaryotic Anabaena are both photosynthetic organisms and their pigmentation is complementary. The Azolla contain chlorophylls a and b as well as carotenoids whereas the Anabaena contain chlorophyll a, phycobiliproteins (35, 51 and Kaplan, unpublished observations) and carotenoids. In A. caroliniana the endophyte accounts for about 15% of the association's chlorophyll and protein (43). The preponderance of Azolla pigments masks the absorption of the Anabaena filaments in the association. Photosynthesis by the isolated endophyte is maximal in the region of phycobiliprotein absorption, but the action spectra for photosynthesis in the Azolla-Anabaena association and endophyte-free Azolla are similar to those of the green plants and virtually indistinguishable from each other (42).

The association and individual partners exhibit Calvin cycle intermediates of photosynthetic CO_2 fixation. Sucrose is a major fixation product in the Azolla, but it is not detectable as a labeled product in the isolated endophyte. While the association and endophyte-free Azolla exhibit an O_2 inhibition of photosynthesis and O_2-dependent CO_2 compensation point, these traits are not exhibited by the isolated endophyte (42).

Attempts to estimate the individual contribution of the Azolla and the Anabaena to the total photosynthetic capability of the association have included differences in temperature sensitivity, action spectra and O_2 inhibition of photosynthesis in the individual partners, as well as $^{14}CO_2$ labeling with rapid isolation of the endophyte (Kaplan, unpublished results). While differences obtained with the association and endophyte-free Azolla with the other procedures had indicated that the endophyte contributes no more than 6-10% of the associations total photosynthetic capability, the $^{14}CO_2$ labeling studies suggest that the endophytes' contribution may be as low as 2%. Moreover, $^{14}CO_2$ labeling followed by isolation of the endophyte after chase periods with air has indicated cross-feeding of fixed carbon from Azolla to the Anabaena (Kaplan and Peters, in preparation).

3.2 Photosynthesis and N_2 Fixation

The close relationship between photosynthesis and N_2 fixation in these associations is well documented (36, 26, 37, 38, 52). Photosynthesis is, of course, the ultimate source of all the ATP and reductant used in N_2 fixation. Dark, aerobic nitrogenase catalyzed reduction of substrates is dependent upon endogenous reserves of photosynthate and dark rates are always less than one-half of those obtained in the light. The cumulative results of a number of studies strongly indicate that: reductant from endogenous reserves of photosynthate and photophosphorylation are the primary driving forces of nitrogenase activity in the light; and that dark, respiratory driven activities may be ATP limited (25, 26, 38).

The interaction between photosynthesis and nitrogenase activity is demonstrated clearly in the action spectra for nitrogenase catalyzed C_2H_2 reduction in the A. caroliniana-Anabaena association and isolated endophyte (52). Although phycobiliproteins are often stated to be associated exclusively with photosystem II-linked processes, the relative rate of C_2H_2 reduction per incident quantum was as great in the region of phycobiliprotein absorption as it was in the region of chlorophyll absorption in both the association and endophyte. Moreover, heterocysts of the endophyte were found to contain phycobiliprotein. More recently, Kaplan et al. (16) have presented information on the total phycobiliprotein as a function of leaf cavity and endophyte age and determined absorption and fluorescence spectra of the phycobiliproteins in individual vegetative cells and heterocysts from leaf cavities of different developmental stages in A. caroliniana and A. pinnata.

3.3 N_2 Fixation and Ammonia Assimilation

Although variations in conditions of prior growth and differences in assay methods limit comparisons of data for different Azolla species, a number of studies have assayed nitrogenase activity using acetylene reduction with rates expressed on the basis of fresh weight, dry weight, total N, chlorophyll or protein (36, 1, 2, 3, 26, 31, 50, 53, 48, 49). Peters et al. determined optimal laboratory culture conditions for A. caroliniana, A. filiculoides, A. mexicana and A. pinnata (39). Comparative physiological studies with these species have recently been completed and preliminary accounts published (33, 34, 15 also see Peters and Ito, this vol, pg. At present the nitrogenase-catalyzed reduction of acetylene, $^{15}N_2$, and protons, and their relationships, are best defined for the A. caroliniana association and the endophyte removed from it (21, 26, 27, 31, 40, 37, 38, 33, 34).

When incubated under $^{15}N_2$ the endophyte isolated from A. caroliniana releases ammonium into the incubation medium (27, 38). These and a related study (43) indicated little or no release of any organic N compounds such as amino acids. Studies of ammonia-assimilating enzymes in the A. caroliniana association and endophyte showed that both partners had glutamine synthetase (GS), glutamate synthase (GOGAT) and glutamate dehydrogenase (GDH) activities. However, the host was estimated to account for at least 90% of the association's GS activities and 80% of the total GDH (43). While low levels of free ammonia can be detected in Azolla (22, 37), incubation of the association under $^{15}N_2$-enriched air followed by chase periods with room air showed a low percentage of the total ^{15}N in the ammonia fraction with a rapid incorporation of ^{15}N into ethanol-soluble compounds and then ethanol-insoluble compounds (37).

4.1 Developmental Physiology

Anabaena filaments associated with the plant apex lack heterocysts and nitrogenase activity. Heterocysts are rapidly differentiated as leaf cavities are occupied by the endophyte. The increase in heterocysts frequency is paralleled by an increase in nitrogenase activity. This developmental profile was described first in A. filiculoides (11). It has been confirmed in another population of A. filiculoides as well as A. pinnata (44) (Kaplan and Peters, unpublished observation) and studied in considerable detail in

A. caroliniana (38, 4, 17). Physiological and biochemical studies of the whole association, endophyte-free plants, and populations of the endophyte isolated from all stages of development, reflect a composite of activities and/or processes. Studies using main stem axes, and individual leaves or segments of the axis bearing sequential groups of leaves, provides a more refined approach to an understanding of structure-function relationships and host-symbiont interactions.

N_2 fixed by the Anabaena in mature leaf cavities is transported to the apical region, meeting the nitrogen requirements of both the plant tissues and the generative Anabaena filaments. Interleaf interaction and the transport of fixed N toward the apex has been demonstrated in main stem axes using a pulse-chase approach with $^{15}N_2$ (17). The transported compound(s) have not been identified and the evidence that N is transferred from the Azolla tissue to the generative Anabaena filaments is indicated. Kaplan and Peters also showed that the N content and dry matter decreased with increasing leaf age, while the C/N ratio increased (17).

These findings are consistent with filaments of the endophyte which actively fix N_2, specifically those in mature cavities, having a diminished capability to metabolize the resulting ammonia (43). They imply also that the host exerts a control or regulation of metabolic processes in the endophyte as a function of the developmental gradient causing the endophyte to rapidly differentiate a disproportionate number of heterocysts and to exist in a state of metabolic idling, serving as an ammonium production facility. The factors responsible for diminished cell division greatly increased heterocysts differentiation, and diminished ability to assimilate the ammonia from N_2 fixation during the developmental profile in the Azolla endophyte are not yet resolved. Other plant-cyanobacterial symbioses do not exhibit a comparable developmental gradient. However, their cyanophytes generally have low or undetectable levels of glutamine synthetase (GS) (46, 9). Low GS levels have been postulated as a biochemical mechanism explaining ammonia excretion (46) and it has been suggested that the host plants might produce effector substances which modify the endophyte's ammonia assimilating pathways by inhibiting its GS activity or synthesis (46, 9).

In A. caroliniana Ray et al. suggested that the endophyte's GS activity might be associated primarily with the undifferentiated filaments (43). Haselkorn and coworkers employed an antibody against the purified GS from Anabaena 7120 and demonstrated that the antigen levels of the endophytic Anabaena are only 5-10% of those observed in a free-living Anabaena and that the antigen concentration was greatest in the endophyte associated with younger leaves (10, 23). Thus, there is reason to suspect that the endophyte's GS decreases as heterocyst frequencies and the population of epidermal hairs in the leaf cavities increase.

Appreciable levels of GDH were found in the Azolla endophyte (43). GDH has an appreciably lower affinity for ammonia than does GS and it is suggested that the endophyte's GDH might be associated with filaments actively fixing N_2. Ammonia released by the endophyte would normally be assimilated by the Azolla GS, but the endophyte's GDH might provide a regulatory role, enabling it to re-assimilate released ammonia at high intra-cavity ammonia concentrations.

The extent of interaction in fern-endophyte carbon metabolism also may vary as a function of the developmental profile. It has been suggested that the endophyte might exhibit photoheterotrophic (35) or mixotrophic (25) metabolism. Ray et al. suggested that sucrose from _Azolla_ might serve as a reduced carbon source for the endophyte in mature leaf cavities with metabolism of sucrose conceivably providing reductant for N_2 fixation (42). Sucrose, glucose and fructose have recently been identified as the major soluble di- and monosaccharides in the endophyte (Peters & Kaplan, 1981) and $^{14}CO_2$ time course studies (Kaplan and Peters, in preparation) indicate that the sucrose found in the endophyte is synthesized by the _Azolla_. Thus, we postulate a transition from photoautotrophic metabolism in generative filaments to a photoheterotrophic or mixotrophic mode of metabolism with increasing differentiation of heterocysts and an exogenous carbon source, sucrose, utilized to maintain levels of reducing power.

5. Current Hypothesis of Functional Organization

The endophyte is always protected by the dorsal leaf lobe of the _Azolla_, and is never directly exposed to the natural aquatic and atmospheric environments. Mineral nutrients and water must be taken up through the fern prior to their reaching the cyanophyte; even the light energy incident on the cyanophyte has been filtered by the fern leaf tissues. In short, the endophytic _Anabaena_ receives physical protection, required minerals and other nutrients, and adequate moisture in a highly specialized environment created for it.

   Our current interpretation of the functional organization of the symbiosis as it relates the interaction of nitrogen and carbon metabolism between the partners is as follows. We hypothesize that the branched hairs are the principal site of nitrogen interchange. Nitrogen fixed in mature cavities is transported to the apical region where terminal cells of the PBH are associated with the apical _Anabaena_ colony. We suggest that the PBH may provide a source of fixed nitrogen and other nutrients to these undifferentiated filaments of the endophyte. As a leaf cavity develops the PBH undergoes morphological and ultrastructural changes and a SBH develops. Both branched hairs are present in leaf cavities where the endophyte has differentiated heterocysts and exhibits nitrogenase activity and the two hairs are located in the same position in each cavity, always on the path of the foliar trace. We therefore postulate that the individual cells of the two branched hairs may have multiple roles as a function of the age of the leaf cavity. These roles would include providing the endophyte with H_2O and mineral nutrients and taking up fixed nitrogen released by the endophyte as ammonium. In the latter role specific cells, perhaps the body cells of the branched hairs, would be expected to contain high levels of ammonia assimilating enzymes, specifically glutamine synthetase. The assimilated N (amino acids) would move through the stalk cell of the branched hairs to the foliar trace and then into the stem stele where they would be transported the apical region. While the simple hairs increase in parallel with increasing heterocyst frequency in the endophyte, they are restricted to those portions of the cavity surrounded by photosynthetic mesophyll. We do not hypothesize any involvement of the simple hairs in the assimilation of ammonium. Rather, we suggest that they may provide a conduit for the transfer of photosynthate, specifically sucrose, from the <u>Azolla</u> to the endophyte. The increase in simple hairs with increasing nitrogenase activity is consistent with this postulate. The <u>Azolla</u> photosynthate metabolized by the endophyte would help satisfy the high demand for reductant which must occur at high heterocyst frequency and high nitrogenase activity.

6. Growth and Physiological Processes as a Function of Photoperiod, and Input from N_2 and Combined N Sources

Using clones of <u>A. caroliniana</u>, <u>A. filiculoides</u>, <u>A. mexicana</u> (two populations) and <u>A. pinnata</u> we have determined biomass increase, C and N content percentage dry matter and chlorophyll a/b ratios as a function of nutrient solution, pH, temperature, photoperiod and light intensity (39). Controlled environment studies were supplemented by glasshouse studies. The tolerance of the individual species to elevated temperature was <u>A. mexicana</u> > <u>A. pinnata</u> > <u>A. caroliniana</u> > <u>A. filiculoides</u>. Using the optimum growth temperature, a 16 hr light period and a photon flux density of at least 400uE/m^2.sec, the <u>Azolla</u> species were shown to double their biomass in 2 days or less and contain 5-6% N on a dry weight basis with N_2 fixed by the endophyte as the only N source.

6.1 <u>Azolla</u> as a Biofertilizer for Rice

While the past and present use of <u>Azolla</u> merits brief comment, it is the future we are concerned with at this workshop. In regard to the past I would simply note that in addition to the work with <u>Azolla</u> as a biofertilizer for rice in Vietnam and China (20, 18, 6, 19) field studies with <u>Azolla</u> were initiated about 6 years ago at a number of institutes, including the International Rice Research Institute (IRRI) in the Philippines (53), the Central Rice Research Institute (CRRI) at Cuttack, India, and the University of California at Davis, California (50). These initial studies assessed the problems and applicability of <u>Azolla</u> and confirmed its potential as an alternative N source for rice. Subsequently, there have been a number of reports by each of these groups as well as a study conducted by Lumpkin in China. I will not attempt to review their findings but would refer interested parties to Singh (45), Talley and Rains, (48, 49), Watanabe et al. (53), and Lumpkin and Plucknett (19) for more recent findings. I would also note that for the past several years IRRI has conducted training programs on <u>Azolla</u> and its utilization with rice. IRRI also has sponsored an international collaborative field trial on the effect of phosphorus on growth, N and P content of <u>Azolla</u>. In 1979 this trial included 14 sites in 5 countries and in 1980 it included 19 sites in countries including Thailand, India, China, Nepal, Bangladesh, Indonesia, India, Sri Lanka and Senegal (13, 14). At present there is no question that <u>Azolla</u> has the capacity to offset N requirements of rice in a number of geographical areas.

Following a special workshop on Nitrogen Fixation and Utilization held at IRRI in April 1980, I wrote the following comments and

and recommendations for research and training on the Azolla-Anabaena symbioses on behalf of the working group.

These photosynthetic associations have certain advantages as an alternative N source for rice. The advantages include the following: (1) they are available, (2) they have demonstrated potential in providing N to rice, (3) they will continue to fix atmospheric N_2 in the presence of fertilizer N as ammonia, nitrate or urea (i.e. their nitrogenase is not appreciably repressed by combined N). Moreover, there is every reason to believe that the use of these associations can be increased. This essentially requires their further development through field and laboratory studies in much the same manner as a new crop plant.

Areas of research and research priorities include:

Management systems: Field application and design, cost analysis, etc.

Revision of classification: It must be demonstrated that there are indeed only six species. Secondly, varietal, strain, or population differences must be determined. It is essential to know what associations various investigators are using. At present, voucher specimens should be deposited at various herbaria for this purpose and in the future it will be necessary to sort out whether populations of individual "species" with different environmental tolerances or nutrition requirements are in fact polyploids or hybrids.

IRRI should continue to serve as a depository and collection center for Azolla. An individual should be in charge of them for distribution to systematists willing to undertake classical as well as chemotaxonomy in order to determine the extent of similarities or differences among populations within a species.

IRRI should provide species or populations with known environmental tolerance to individuals conducting field studies. Other species or populations may be superior to the endemic one in regard to temperature tolerance, N_2 fixing capability, pest resistance, etc.

Screening of species and strains for those efficiently utilizing phosphate should be continued. Preloading should be studied and approaches to providing phosphate in the form of a slow release encapsulated material explored to maximize utilization by the Azolla and minimize loss.

N balance studies are needed and additional information is required on the rates of mineralization and N losses through denitrification, etc.

Physiological studies of photosynthesis, photorespiration, N_2 fixation, H_2 production, etc. would be desirable under field conditions.

Studies are needed on how to induce or inhibit sporulation. If control of sporulation is demonstrated, methods of storage of spores as a germplasm should be determined along with whether there is any dormancy involved after prolonged storage. Tissue culture and cold storage should also be studied as a means of preserving stocks. (Cold storage refers to growth at minimal temperature to minimize need of transfer, but it must also be shown that prolonged treatment under such conditions does not alter environmental tolerances).

Diseases along with insect and other pests of Azolla should be delineated and the control pursued.

Effects of herbicides including residual effects, on Azolla should be determined.

Basic morphological, physiological and biochemical studies of promising species or strains should be pursued under laboratory conditions to further understand the working of these associations, including the effects of sporulation on physiological process. Eventually, attempts to create crosses or hybrids with specifically desired traits should be undertaken".

It is my opinion, as it was then, that these constitute the immediate areas of research requiring attention. A few additional points are in order. The use of Azolla is labor intensive and as yet I am not aware of anyone conducting a cost analysis comparing the manpower expended with Azolla versus the cost of using an equivalent amount of commercial fertilizer. In regard to the labor intensiveness it is worth noting that thought has been given to mechanization of some aspects of Azolla use. Next, I would note that the phosphorus requirements of Azolla may be overdramatized. It is important to realize that the phosphate requirement appears to vary with different strains or populations within an Azolla species (47, Talley et al., 1981; and Talley, personal communication) and that it may well vary with local management practices. The reason for the often noted P requirement is that the phosphate rapidly gets tied up in the paddy soil and is unavailable to the floating Azolla. However, the phosphate in the paddy soil is available to the rice plants and so is the phosphate taken up by Azolla when the latter is incorporated (48, 49) into paddy soil.

An understanding of those factors which control sporulation – i.e., cause as well as prevent the process is essential to the future practical use of Azolla. Details of the Azolla life cycle have been presented previously (19, 28, 29) and scanning electron micrographs of the reproductive structures from A. mexicana are included in these proceedings (Calvert et al.). Azolla is amazing in that it is the only symbiosis in which the prokaryotic partner is carried through the sexual cycle of the eukaryote.

Azolla spores should provide a means of storing desired germplasms and, if obtainable in sufficient quantity, may have potential as an inoculum. The N input of Azolla is a function of biomass and it makes little difference whether one uses an inoculum of living material or inoculates with spores so long as an adequate biomass is achieved. A large inoculum of spores may eventually be an alternative, and less labor intensive approach, than the use of sporophytes.

Production of spores should also open the way to breeding programs in which intra- and inter-specific crosses may enable us to obtain "super" Azolla with specifically desired traits and form the beginning of Azolla genetics. Although it is not yet known whether endophyte-free Azolla can be induced to sporulate, the spores from such plants would provide a possible means of introducing new, genetically-improved endophytes. In short, if sporulation can be induced, a method developed for harvesting the spores, and the spores stored for prolonged periods, a number of current obstacles may be overcome and new vistas opened. However, it is also important to be able to prevent sporulation since when it occurs the rates of growth and N_2 fixation in the field have been reported to decline (49).

Acknowledgement

I am deeply indebted to a number of my current and former associates for their contributions to the _Azolla_ research conducted at the C.F. Kettering Research Laboratory. These include: Drs. H.E. Calvert, W.R. Evans, O. Ito, D. Kaplan, B.C. Mayne, V.V.S. Tyagi and T.B. Ray, Ms. D.K. Crist, R.E. Poole and S.K. Perkins, Mr. M.K. Pence, and especially Mr. R.E. Toia, Jr.

References

1 Ashton P J and Walmsley R D 1976 The aquatic fern _Azolla_ and its _Anabaena_ symbiont. Endeavour 35; 39–43.

2 Becking J H 1976 Contributions of plant algal associations. In: W E Newton and C J Nyman eds. Proceedings of the 1st International Symposium on Nitrogen Fixation Vol. II. Washington State University Press. Pullman, Washington. pp. 581–591.

3 Becking J H 1979 Environmental requirements of _Azolla_ for use in tropical rice production. In: Nitrogen and Rice. International Rice Research Institute, Los Banos, Laguna, Philippines. pp. 345–374.

4 Calvert H E and Peters G A 1981 The _Azolla-Anabaena azollae_ relationship. IX. Morphological analysis of leaf cavity hair populations. New Phytol. 89; 327–335.

5 Calvert H E and Peters G A 1982 A scanning electron microscopic view of sporulation in _Azolla-mexicana_ Presl. (these proceedings).

6 Dao T T and Tran Q T 1979 Use of _Azolla_ in rice production in Vietnam. In: Nitrogen and Rice. International Rice Research Institute, Los Banos, Laguna, Philippines. pp. 395–405.

7 Duckett J G, Toth R and Soni S L 1975 An ultrastructural study of the _Azolla_, _Anabaena azollae_ relationship. New Phytol. 75; 111–118.

8 Gunning B E S, Hughes J E and Hardham A R 1978 Formative and proliferative cell divisions, cell differentiation, and developmental changes in the meristem of _Azolla_ roots. Planta 143; 121–144.

9 Haselkorn R 1978 Heterocysts Annu. Rev. Plant Physiol. 29; 319–344.

10 Haselkorn R, Mazur B, Orr J, Rice D, Wood N and Rippka R 1980 Heterocyst differentiation and nitrogen fixation in cyanobacteria (blue-green algae). In: W E Newton and W H Orme-Johnson, eds. Nitrogen Fixation, Vol. II. University Park Press, Baltimore. pp. 259–278.

11 Hill D J 1975 The pattern of developmental of _Anabaena_ in the _Azolla-Anabaena_ symbiosis. Planta 133; 237–242.

12 Hill D J 1977 The role of _Anabaena_ in the _Azolla-Anabaena_ symbiosis. New Phytol. 78; 611–616.

13 International Rice Research Institute 1980 Report on the first trials of _Azolla_ use to rice, INSFFER (1980). IRRI, Los Banos, Laguna, Philippines.

14 International Rice Research Institute 1981 Report on the second

trials of Azolla use to rice, INSFFER (1980). IRRI, Los Banos, Laguna, Philippines.

15   Ito O, Toia R E, Jr., Poole R E, Crist D K, Evans W R, Mayne B C and Peters G A   1980   Physiological studies on N_2-fixing Azolla species grown under three photoperiods. Plant Physiol. 65; S-109.

16   Kaplan D, Calvert H E and Peters G A   1982   Phycobiliprotein in the Azolla endophyte as a function of leaf age and cell type. Plant Physiol. 69; S-156.

17   Kaplan D and Peters G A   1981   The Azolla-Anabaena azollae relationship.   X. $^{15}N_2$ fixation and transport in main stem axes. New Phytol. 89; 337-346.

18   Liu C C   1979   Use of Azolla in rice production in China. In: Nitrogen and Rice.   International Rice Research Institute, Los Banos, Laguna, Philippines. pp. 375-394.

19   Lumpkin T A and Plucknett D L   1980   Azolla: botany, physiology, and use as a green manure. Econ. Bot. 34; 111-153.

20   Moore A W   1969   Azolla: biology and agronomic significance. Bot. Rev. 35; 17-34.

21   Newton J W   1976   Photoproduction of molecular hydrogen by a plant algal symbiotic system. Science 191; 559-561.

22   Newton J W and Cavins J F   1976 Altered nitrogenous pools induced by the Azolla-Anabaena symbiosis.   Plant Physiol. 58; 798-799.

23   Orr J and Haselkorn R   1982   Regulation of glutamine synthetase activity and synthesis in free-living Anabaena and in symbiotic association. J. Bacteriol. 152; 626-635.

24   Pate J S and Gunning B E S   1972   Transfer cells. Annu. Rev. Plant Physiol. 23; 173-196.

25   Petes G A   1975   The Azolla-Anabaena azollae relationship. III. Studies on metabolic capabilities and a further characterization of the symbiont. Arch. Microbiol. 103; 113-122.

26   Peters G A   1976   Studies on the Azolla-Anabaena azollae symbiosis.   In: W E Newton and C J Nyman, eds.   Proceedings of the 1st International Symposium on Nitrogen Fixation, Vol. II. Washington State University Press, Pullman, Washington. pp. 592-610.

27   Peters G A   1977   The Azolla-Anabaena azollae symbiosis. In: A. Hollaender, ed.   Genetic Engineering for Nitrogen Fixation. Plenum Press, New York. pp. 231-258.

28a Peters G A and Kaplan D   1981   Soluble carbohydrate pool in the Azolla-Anabaena symbiosis. Plant Physiol. 67; S-37.

28b Peters G A and Calvert H E   1982a The Azolla-Anabaena symbioses. In: N S Subba Rao, ed.   Advances in Agricultural Microbiology. Oxford and IBH Publ. Co., New Delhi. pp. 191-218.

29   Peters G A and Calvert H E   1982b   The Azolla-Anabaena azollae symbioses.   In: Lynda Goff, ed.   Algal Symbiosis: Cambridge University Press, New York. pp. 109-145.

30   Peters G A, Calvert H E, Kaplan D, Ito O and Toia R E, Jr. 1982 The Azolla-Anabaena symbiosis. Israel J. Bot. 30.

31   Peters G A, Evans W R and Toia R E, Jr.   1976   Azolla-Anabaena azollae relationship.   IV. Photosynthetically-driven nitrogenase-catalyzed H_2 production. Plant Physiol. 58; 119-126.

32   Peters G A and Ito O   1982   Determining N_2 fixation and N input in Azolla, grown with and without combined nitrogen sources: Keeping the acetylene reduction assay in the proper perspective. (these proceedings).

33 Peters G A, Ito O, Tyagi V V S and Kaplan D 1981a Physiological studies on N_2-fixing Azolla. In: J M Lyons, et al., eds. Genetic Engineering of Symbiotic Nitrogen Fixation and Conservation of Fixed Nitrogen. Plenum Press, New York. pp. 343-362.

34 Peters G A, Ito O, Tyagi V V S, Mayne B C, Kaplan D and Calvert H E 1981b Photosynthesis and N_2 fixation in the Azolla-Anabaena symbiosis. In: A H Gibson and W E Newton, eds. Current Perspectives in Nitrogen Fixation. Australian Academy of Science, Canberra. pp. 121-124.

35 Peters G A and Mayne B C 1974a The Azolla, Anabaena azollae relationship. I. Initial characterization of the association. Plant Physiol. 53; 813-819.

36 Peters G A and Mayne B C 1974b The Azolla, Anabaena azollae relationship. II. Localization of nitrogenase activity as assayed by acetylene reduction. Plant Physiol. 53; 820-824.

37 Peters G A, Mayne B C, Ray T B and Toia R E, Jr. 1979 Physiology and biochemistry of the Azolla-Anabaena symbiosis. In: Nitrogen and Rice. International Rice Research Institute, Los Banos, Laguna, Philippines. pp. 325-344.

38 Peters G A, Ray T B, Mayne B C and Toia R E Jr. 1980a Azolla-Anabaena association: morphological and physiological studies. In: W E Newton and W H Orme-Johnson, eds. Nitrogen Fixation Vol. II. University Park Press, Baltimore. pp. 293-309.

39 Peters G A, Toia R E Jr., Evans W R, Crist D K, Mayne B C and Poole R E 1980b Characterization and comparisons of five N_2-fixing Azolla-Anabaena associations. I. Optimization of growth conditions for biomass increase and N content in a controlled environment. Plant, Cell and Environ. 3; 261-269.

40 Peters G A, Toia R E Jr. and Lough S M 1977 Azolla-Anabaena azollae relationship. V. $^{15}N_2$ fixation, acetylene reduction, and H_2 production. Plant Physiol. 59; 1021-1025.

41 Peters G A, Toia R E Jr., Raveed D and Levine N J 1978 The Azolla-Anabaena azollae relationship. VI. Morphological aspects of the association. New Phytol. 80; 583-593.

42 Ray T B, Mayne B C, Toia R E Jr. and Peters G A 1979 Azolla-Anabaena relationship. VIII. Photosynthetic characterization of the association and individual partners. Plant Physiol. 64; 791-795.

43 Ray T B, Peters G A, Toia R E Jr. and Mayne B C 1978 Azolla-Anabaena relationship. VII. Distribution of ammonia-assimilating enzymes, protein, and chlorophyll between host and symbiont. Plant Physiol. 62; 463-467.

44 Shi D J, Li J G, Zhung Z P, Wang F Z, Zhu L P and Peters G A 1981 Studies on nitrogen fixation and photosynthesis in Azolla imbricata (Roxb.) and Azolla filiculoides Lam. Acta Bot. Sin. 23; 306-315 (in Chinese).

45 Singh P K 1980 Introduction of "Green Azolla" biofertilizer in India. Curr. Sci. 49; 155-156.

46 Stewart W D P 1977 A botanical ramble among the blue-green algae. Br. Phycol. J. 12; 89-115.

47 Subudhi B P R and Watanabe I 1981 Differential phosphorus requirement of Azolla species and strains in phosphorus limited continuous culture. Soil Sci. Plant Nutr. 27; 237-247.

48 Talley S N and Rains D W 1980a Azolla as a nitrogen source for temperate rice. In: W E Newton and W H Orme-Johnson eds.

Nitrogen Fixation Vol. II. University Park Press, Baltimore.
pp. 311–320.

49 Talley S N and Rains D W 1980b _Azolla_ _filiculoides_ Lam. as
a fallow season green manure for rice in temperate climate.
Agron. J. 72; 11–18.

50 Talley S N, Talley B J and Rains D W 1977 Nitrogen fixation
by _Azolla_ in rice fields. _In_: A. Hollaender ed. Genetic Engineering for Nitrogen Fixation. Plenum Press, New York. pp. 259–
281.

51 Tyagi V V S, Mayne B C and Peters G A 1980 Purification and
initial characterization of phycobiliproteins from the endophytic
cyanobacterium of _Azolla_. Arch. Microbiol. 128; 41–44.

52 Tyagi V V S, Ray T B, Mayne B C and Peters G A 1981 The _Azolla-
Anabaena azollae_ relationship. XI. Phycobiliproteins in the action
spectrum for nitrogenase–catalyzed acetylene reduction. Plant
Physiol. 68; 1479–1484.

53 Watanabe I, Espinas C R, Berja N S and Alimagno B V 1977 Utilization of the _Azolla-Anabaena_ complex as a nitrogen fertilizer
for rice. Int. Rice Res. Inst. Res. Paper Ser. 11; 1–15.

FIRST SECTION: BASIC STUDIES

2.   Formation and Breakdown of <u>Azolla</u>   ypeptides in Paddy Soils

J.W. Newton and D.D. Tyler
Northern Regional Research Center
Agricultural Research Service
U.S. Department of Agriculture
Peoria, Illinois 61604

Key words   <u>Azolla</u> CGP   Cyanobacteri   Cyanophycin   Polypeptides
SDS

Summary

The total polypeptide patterns of Azolla plants and symbiotic algae (cyanobacteria) grown on $^{14}CO_2$ were examined by isoelectric focusing SDS polyacrylamide gel electrophoresis.   The number of resolved polypeptides was greater in algal than in plant extracts. A component always found in dinitrogen-fixing Azolla was multiarginine-polyaspartic acid (cyanophycin), a reserve polymer of algal akinetes. This finding indicated that a substantial fraction of Azolla nitrogen is reserve polypeptide derived from senescent forms of the symbiont. ^{14}C-Labeled cyanophycin was rapidly broken down in paddy soils with release of $^{14}CO_2$.   Hence, although insoluble and generally resistant to proteolytic digestion, cyanophycin could serve as a major source of available nitrogen in paddy soils treated with Azolla.

1. Introduction

The suitability of Azolla for use as a green manure for wetland agriculture in both tropical and temperate regions is now well documented (7, 8).   The response of plants to N added as Azolla is rapid, and it can completely substitute for inorganic N at levels below about 40 kg/ha (26).   These results indicate that much of the Azolla nitrogen is rapidly made available for plant growth. Furthermore, Azolla is unusual in composition since approximately 15% of its protein is derived from the blue-green algal (cyanobacterial) symbiont (17) and therefore is of microbial origin. We have examined the proteins present in Azolla and its symbiont, using electrophoretic-radiographic techniques capable of resolving and detecting nanogram amounts of any cell protein.   These studies, as well as studies on the relative stability of some cyanobacterial proteins in paddy soils, are reported in this communication.

2. Materials and Methods

<u>Azolla</u> <u>caroliniana</u> and a derived cyanobacterial-free strain were grown as described by Peters and Mayne (13). Cyanobacteria were grown on the BG-11 medium of Stanier et al. (24), without nitrate. To obtain uniformly labeled protein, plants and cyanobacteria were exposed to about 0.2 mC $^{14}CO_2$ for 24-48 hour; generally, one to two plants or 5-10 ml algal cultures containing about 1 mg protein

were labeled. Cyanobacterial bundle preparations were isolated from enzymatic digests of labeled Azolla by the method of Newton and Herman (9). The samples were macerated in a lysis buffer containing ampholines, 9 $\underline{M}$ urea and mercaptoethanol, subjected to 2 min sonic disruption, frozen, thawed, and then centrifuged. The general procedure for preparation and separation of labeled proteins was that of O'Farrell (12). Isoelectric focusing in the first dimension (pH range 4-8) was followed by electrophoresis on nongradient 15% acrylamide gels containing sodium dodecyl sulfate (SDS). This procedure separates proteins first on the basis of their isoelectric points and, in the second dimension, based on their molecular weight. The extracts contained approximately 5 micrograms protein and 10^6 counts/min ^{14}C applied to the gels. Following separation, fluorography of the radioactive gels (2) allowed detection of about 1 nanogram of a protein or 10^2 counts/min ^{14}C in a single protein spot.

Cyanophycin granule polypeptide (CGP) was isolated from cyanobacteria and assayed by the method of Simon (19). $\underline{Anabaena}$ strain 2B, originally isolated from $\underline{A}$. $\underline{caroliniana}$ (9) was used for most experiments. Another strain, 10B, characterized as a $\underline{Nostoc}$ sp. and isolated from paddy soils in which Azolla were abundant, was also used to prepare ^{14}C labeled CGP, because it forms large numbers of akinetes (Newton, in press 1982).

For assays of CGP breakdown, soil samples were diluted with water to approximately 5 mg dry weight per ml and, if necessary, adjusted to pH 7 with phosphate or tris buffers. Above this concentration, the breakdown of CGP, which appeared confined to soil particulate matter, was not linear with increasing amounts of soil added to reaction mixtures, making comparisons unreliable. To measure $^{14}CO_2$ release, a small glass rod, etched on one end, was dipped in 1 N NaOH and inserted above the reaction mixture through a hole in a rubber stopper. The $^{14}CO_2$ collected on the rod was counted by rinsing the alkali into vials with a liquid scintillation counting solution. Acidification of reaction mixtures after incubation showed that recovery of $^{14}CO_2$ from the solutions was essentially complete during incubation when a 2.0-ml volume was used in a closed 15-ml Erlenmeyer flask.

Soil samples were obtained with the cooperation of Drs. D. Bartholomew and M. Habte, Department of Agronomy and Soil Science, University of Hawaii at Manoa, and were collected in Hasu (lotus) and Taro paddies in which Azolla, although not specifically being used as a green manure, was naturally abundant.

3. Results

Figure 1a is a typical fluorograph of a pattern obtained on a two-dimensional gel of $\underline{Azolla}$ $\underline{caroliniana}$ polypeptides from plants grown on nitrogen-free media and containing cyanobacterial symbionts. The range of isoelectric points of the separated polypeptides from isoelectric focusing is indicated at the top of the figure. The molecular weight range (approximately 200-10,000) of components separated in the second dimension by polyacrylamide-sodium dodecyl sulfate electrophoresis of the focused components is also indicated. The gels were calibrated using ^{14}C-labeled protein standards of known molecular weights, as indicated on the figure. For comparison, Figure 1b is a fluorograph of a gel prepared using an extract from the cyanobacterial-free strain of A. caroliniana grown on media containing nitrate, showing a much smaller number of components. Figure 1c is a gel run using a polypeptide extract from ^{14}C-labeled Anabaena. Several features are apparent from the gel patterns. First, the method resolves microbial proteins much better than those derived from plant tissue. The number of polypeptides resolved in the extract from dinitrogen-fixing Azolla, which contains cyanobacterial symbionts, is also larger than that from cyanobacteria-free plants, and many of these polypeptides are of algal origin. A cluster of spots having isoelectric points at about 5.5 and molecular weights in the 50-thousand range is observed in dinitrogen fixing plants, as well as a band of spots at I.P. 7 and 50-200,000 MW; a broad band at I.P. 4.75 is also observed, which is characteristic of cyanobacterial material. Furthermore, in contrast to cyanobacterial preparations, which have a wide variety and number of protein components, the plant proteins separated generally have acidic isoelectric points, and substantial amounts appear to be high-molecular-weight material which is not migrating during electrophoresis. Several attempts to increase resolution of material in the plant extracts by treatment with nucleases, lipid removal, and variations in the procedure were unsuccessful. It is obvious that the method is not resolving all of the plant proteins. Similar results have been obtained by other investigators who have used the method on plant material (3). It is probable that as more investigators apply these methods to various plant extracts, improvements in the procedures will evolve. The O'Farrell method was originally developed using extracts from microorganisms and is even capable of detecting mutational changes in polypeptides from such sources. Hence, its resolving power is very high, and the inability to demonstrate a large number of proteins in the plant extracts may be due to a combination of a larger amount of structural material as well as complex formation in the plant extracts. However, the polypeptide patterns obtained are characteristic of the materials from which they are derived; by simple examination of the fluorogram, one can tell from the pattern the source of the polypeptides from which it was made.

When cyanobacterial bundle preparations were isolated from enzymatic digests of ^{14}C-labeled A. caroliniana, the polypeptide pattern obtained was also characteristic of the material. As illustrated in Figure 2a, the cluster of cyanobacterial proteins is present; in addition, the large band at I.P. 4.75 appears on the fluorogram and runs down the gel upon electrophoresis, covering a molecular weight range of 100-20 thousand. This material is characteristic of bundle preparations from Azolla and also becomes more prominent in older cultures of Anabaena sp. grown on N free media, as seen in Figure 2b. The component has been identified by composition and solubilities as cyanophycin granule polypeptide or CGP, the characteristic storage material of cyanobacterial akinetes (18, 22). Figure 2c is a fluorogram of CGP isolated from ^{14}C-labeled cyanobacteria run on a standard two-dimensional gel. The electrophoretic properties of this polypeptide, including isoelectric point (4.75) and molecular weight range (25-100,000), agree closely with those reported by Simon and Weathers (22) in their characterization of this material.

It is apparent from these results that, when Azolla is used as a green manure, considerable amounts of CGP could be incorporated into paddy soils. CGP is an insoluble, branched, 1:1 copolymer of aspartic acid and arginine that is highly resistant to enzymatic breakdown. Consequently, we examined the stability of this polypeptide in paddy soils, using purified ^{14}C-labeled CGP prepared from algal cultures grown on [14]CO_2. Since CGP is insoluble at neutral pH but soluble in dilute acid (19), several assays were developed to study its possible breakdown, based on centrifugation of reaction mixtures after incubation of ^{14}C-CGP with soil samples. Preliminary studies revealed that, instead of simply being hydrolyzed in soils, ^{14}C was lost from CGP. This result led us to measure [14]CO_2 released from the samples as a sensitive measure of CGP breakdown in paddy soils. Table 1 illustrates the results of a typical experiment in which labeled CGP was added to a paddy soil at neutral pH. Because CGP is insoluble, centrifugation of the reaction mixture after incubation would remove labeled substrate. In practice, at the low concentration of [14]CGP used, addition of several volumes of ethanol is also necessary to remove all radioactivity. Acidification of a reaction mixture sample to solubilize CGP, followed by centrifugation, allows measurement of total CGP present. By using appropriate zero time and sterilized soil controls, one can correct for adsorption of CGP to soil particles and possible nonenzymic breakdown. As indicated in Table 1, approximately 10% of the added ^{14}C was lost as [14]CO_2 in 24 hours.

Data in Table 2 show the time course of [14]CO_2 release, which was linear for up to 8 hours, and indicate that sterile (autoclaved) soil was inactive.

It might be expected that breakdown of CGP would be more extensive in paddy soils containing Azolla, since such soils would contain cyanobacteria and CGP in abundance. However, an initial survey indicated that all soils tested could release [14]CO_2 from CGP. The examples cited in Table 3 all show substantial activity in release of [14]CO_2; differences noted among soils are probably not significant.

An estimate of the relative breakdown of CGP compared to other cyanobacterial proteins was then made by use of crude sonic extracts of ^{14}C labeled cyanobacteria, which were dialyzed extensively to remove ^{14}C-labeled material of low molecular weight. The extract was centrifuged at 25,000 g for 20 min to remove CGP as well as other insoluble constituents, and ^{14}C-labeled CGP was isolated from the extract for assay. Equal amounts of radioactivity from these fractions were then added to paddy soil samples, and the [14]CO_2 released was measured after 24 hours. The data in Table 4 show that the purified CGP sample yielded approximately 30% as much [14]CO_2 as the total cyanobacterial extract upon incubation with the paddy soil sample. The crude dialyzed extract contains all cell components, many of which could be presumed to be readily attacked by soil microorganisms. It appears, therefore, that CGP is broken down by the soil microflora at a rate comparable to that of any other cell constituent.

4. Discussion

Leaf development and cyanobacterial grown in Azolla is a coordinated process (4, 5, 14, 15), with nitrogenase activity declining in older leaves, accompanied by an increase in cell size and granulation of the algae within the leaf cavity. These are all properties of cyanobacterial akinetes (11, 21, 25) which, although not "spores" in the usual sense attributed to microorganisms, are analogous resistant cells associated with reproduction and perennation. Akinetes also contain large amounts of cyanophycin granule polypeptide, CGP a reserve polymer (18) the formation of which accompanies akinete development in older cultures (19, 25); Newton, in press, 1982). These findings indicate that CGP in Azolla is associated with akinete formation within the leaf cavity and could result from restrictive growth conditions in older leaves. Since CGP can constitute as much as 10% of the dry weight of cyanobacterial cells (1, 19), which constitute from 10-20% of the plant protein, CGP itself could represent several percent of the total material incorporated into paddies as green manure when Azolla is used.

   CGP is a branched, 1:1 copolymer of aspartic acid and arginine and hence contains approximately 25% nitrogen. CGP was found by Simon and Weathers to be resistant to digestion by such proteolytic enzymes as trypsin, pepsin, pronase, leucine aminopeptidase, and carboxypeptidases A and B (23). The polymer is formed by a non-ribosomal system (20) and is degraded only upon germination of akinetes. It, therefore, appears to be a nitrogen reserve in cyanobacteria, and might be considered to be the microbial analog of arginine-rich storage proteins often found in plant seeds. Such seeds rapidly degrade arginine upon germination (6, 23) with release of urea and carbon dioxide. The rapid breakdown of ^{14}C-CGP in paddies, with release of $^{14}CO_2$, suggests that the polymer may follow a similar path in providing an ideal nutrient for initiation of plant growth.

   Finally, it is worth noting that use of Azolla for plant production results in repeated inoculation of paddies with cyanobacteria capable of both dinitrogen fixation and CGP production. Because this practice would enhance the cyanobacterial population of the paddy, CGP could eventually play an important role in nitrogen turnover and storage in paddy soils, apart from its initial presence in Azolla.

References

1 Allen M M, Hutchison F and Weathers P J 1980 Cyanophycin granule polypeptide formation and degradation in the cyanobacterium Aphanocapsa 6308. J. Bacteriol. 141: 687-693.

2 Bonner W M and Laskey R A 1974 A film detection method for tritium-labeled proteins and nucleic acids in polyacrylamide gels. Eur. J. Biochem. 46: 83-88.

3 Gilbert C W and Buetow D E 1981 Gel electrophoresis of chloroplast polypeptides: Comparison of one-dimensional and two-dimensional gel analyses of chloroplast polypeptides from Euglena gracilis. Plant Physiol. 67: 623-628.

4 Hill D J 1975 The pattern of development of Anabaena in the Azolla-Anabaena symbiosis. Planta (Berl.) 122: 179-184.

5 Hill D J 1977 The role of Anabaena in the Azolla-Anabaena symbiosis. New Phytol. 78: 611-616.

6 Jones V M and Boulter D 1968 Arginine metabolism in germinating seeds of some members of the leguminosae. New Phytol. 67: 925-934.

7 Lumpkin T A and Plucknett D L 1980 Azolla: Botany, physiology, and use as a green manure. Econ. Bot. 34(2): 111-153.

8 Moore A W 1969 Azolla: Biology and agronomic significance. Bot. Rev. 35: 17-34.

9 Newton J W and Herman A I 1979 Isolation of cyanobacteria from the aquatic fern, Azolla. Arch. Microbiol. 120: 161-165.

10 Newton J W 1982 Manuscript in press.

11 Nichols J M and Carr N G 1978 Akinetes of cyanobacteria. In G. Chambliss and J. C. Vary (eds.), Spores VII. Am. Soc. Microbiol., Washington, D.C., USA, p. 335-343.

12 O'Farrell P H 1975 High resolution two-dimensional electrophoresis of proteins. J. Biol. Chem. 250: 4007-4021.

13 Peters G A and Mayne B 1974 The Azolla Anabaena-Azolla relationship I. Initial characterization of the relationship. Plant Physiol. 53: 813-819.

14 Peters G A, Ray T B, Mayne B C and Toia R E 1980 Azolla-Anabaena Association: Morphological and physiological studies. In W E Newton and W H Orme-Johnson (eds.), Nitrogen Fixation Vol. II, University Park Press, Baltimore, USA, p. 293-309.

15 Peters G A, Ito O, Tyagi V V S and Kaplan D 1981 Physiological studies on N_2-fixing Azolla. In J. M. Lyons, R. C. Valentine, D. A. Phillips, D. W. Rains and R. C. Huffaker (eds.), Genetic engineering of symbiotic nitrogen fixation and conservation of fixed nitrogen, Plenum Publishing Co., New York, USA, p. 343-362.

16 Ray T B, Peters G A, Toia R E and Mayne B C 1978 Azolla Anabaena Relationship VII. Distribution of ammonia assimilating enzymes, protein, and chlorophyll between host and symbiont. Plant Physiol. 62: 463-467.

17 Simon R D 1971 Cyanophycin granules from the blue-green algae Anabaena cylindrica: A reserve material consisting of copolymers of aspartic acid and arginine. Proc. Nat. Acad. Sci. USA 68: 265-267.

18 Simon R D 1973a Measurement of the cyanophycin granule polypeptide in the blue-green algae Anabaena cylindrica. J. Bacteriol. 114: 1213-1216.

19 Simon R D 1973b The effect of chloramphenical on the production of cyanophycin granule polypeptide in the blue-green algae Anabaena cylindrica. Arch. Microbiol. 92: 115-122.

20 Simon R D 1977 Sporulation in the filamentous cyanobacterium Anabaena cylindrica. The course of spore formation. Arch. Microbiol. 111: 283-288.

21 Simon R D and Weathers P 1976 Determination of the structure of the novel polypeptide containing aspartic acid and arginine which is found in cyanobacteria. Biochim. Biophys. Acta 420: 165-176.

22 Splittstoesser W E 1969 Metabolism of arginine by aging and 7 day old pumpkin seedlings. Plant Physiol. 44: 361-366.

23 Stanier R Y, Kunisawa R, Mandel M and Cohen-Bazire G 1971 Purification and properties of the unicellular blue-green algae (Order Chrococcales). Bacteriol. Rev. 35: 171-205.

24 Sutherland J M, Herdman M and Stewart W D P 1979 Akinetes of the cyanobacterium Nostoc PCC 7524: Macromolecular composition, structure, and control of differentiation. J. Gen. Microbiol. 115: 273-287.

25 Talley S N and Rains D W 1980 Azolla filiculoides Lam. as a
   follow-season green manure for rice in a temperate climate.
   Agron. J. 72: 11-18.

Table 1. Breakdown of cyanophycin granule polypeptide (CGP) in paddy soil.

	^{14}C, total counts/min
CGP added	200,000
^{14}C loss	21,160
Recovered as $^{14}CO_2$	20,000

Reaction mixture contained Hasu paddy soil, 12 mg dry weight, and 3 μgm ^{14}C-CGP containing 2×10^5 cpm in 2 ml 0.02 M tris buffer pH 7.5, incubated 24 hours, 25 C.

Table 2. Time course of release of $^{14}CO_2$ from CGP in paddy soil.

	Total $^{14}CO_2$ released, counts/min
Reagent blank	700
Sterile soil	750
Paddy soil	
2 hour	3,870
4 hour	6,660
8 hour	11,010
24 hour	23,050

Reaction mixture contained Hasu paddy soil, 12 mg dry wt, and 3 μgm ^{14}C-CGP, 2 X 10^5 cpm in 2.0 ml 0.02 $\underline{M}$ tris buffer pH 7.5.

Table 3. Comparison of soils for CGP breakdown.

Soil sample	Total $^{14}CO_2$ released, counts/min
Hasu paddy	27,400
Taro paddy	17,040
Cane field	20,400
Garden	8,900

Soil samples were adjusted to contain approximately 10 mg dry weight, incubation 24 hours with 3 μgm ^{14}C-CGP, 2 X 10^5 counts/min in 0.02 M tris 7.5, 25 C.

Table 4. Comparison of breakdown of algal fractions in soil.

	Total $^{14}CO_2$ released, counts/min
Total _Anabaena_ extract	52,000
Soluble fraction after CGP centrifugation	47,200
Isolated CGP	16,280

Hasu paddy soil, 12 mg dry weight, incubation 24 hours; labeled substrates each contained 10^5 counts/min, 0.02 $\underline{M}$ tris 7.5.

28

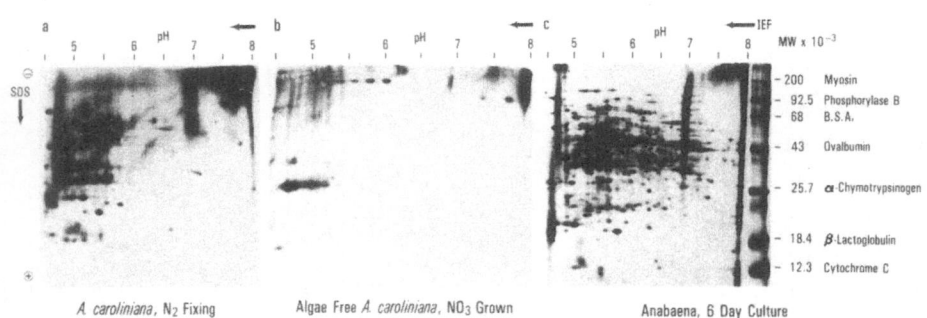

A. caroliniana, N₂ Fixing   Algae Free A. caroliniana, NO₃ Grown   Anabaena, 6 Day Culture

Fig. 1. Fluorographs of two-dimensional electrophoretic gel separations of total polypeptides: (a) <u>A. caroliniana</u> grown on N free medium; (b) <u>A. caroliniana</u>, algal-free strain grown on nitrate; (c) <u>Anabaena</u> strain 2b grown in BG-11 nitrogen free medium. Approximately 5 μgm protein containing 1×10^6 cpm was separated in each preparation.

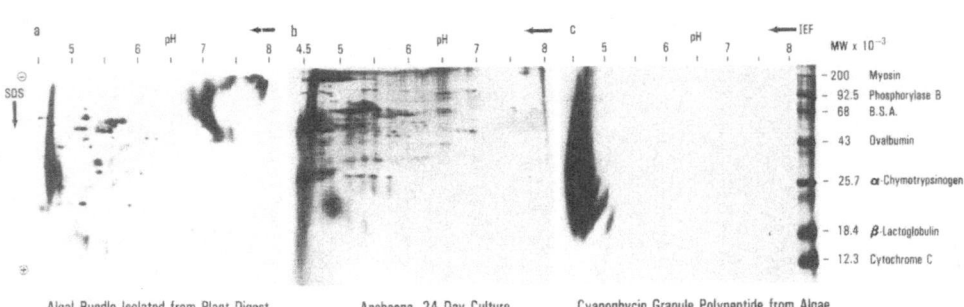

Algal Bundle Isolated from Plant Digest   Anabaena, 24 Day Culture   Cyanophycin Granule Polypeptide from Algae

Fig. 2. Fluorographs of polypeptides separated by two-dimensional gel electrophoresis: (a) cyanobacterial "bundle" preparation isolated from enzymatic digest of Azolla; (b) <u>Anabaena</u>, 24-day culture; (c) purified cyanophycin granule polypeptide.

3. Determining N_2 Fixation and N Input in <u>Azolla</u> Grown with and without Combined Nitrogen Sources: Keeping the Acetylene Reduction Assay in the Proper Perspective

Gerald A. Peters
Charles F. Kettering Research Laboratory
Yellow Springs, Ohio 45387 USA

Osamu Ito
National Institute for Environmental Studies
Ibaraki, Japan

Key words Acetylene Reduction <u>Anabaena</u> <u>Azolla</u> Conversion Factors Nitrogenase Activity

Summary

In the absence of a combined N source in the aqueous environment, N_2 fixation by the endophytic <u>Anabaena</u> can provide <u>Azolla-Anabaena azollae</u> symbioses with their total N requirement. When combined N sorces such as ammonium, nitrate or urea are present, these associations can assimilate the exogenous N sources and fix atmospheric N_2 simultaneously.

The C_2H_2 reduction assay is often used to estimate N_2 fixation, in conjunction with theoretically or experimentally determined conversion factors, and subsequently extrapolated to N fixed per annum. Such approaches ignore the facts that the C_2H_2 reduction assay measures nitrogenase activity at a single point in time, that this activity fluctuates within and between light-dark cycles, and that the experimentally determined conversion factor (C_2H_2 reduced/N_2 fixed) seldom coincides with the theoretical. These aspects are illustrated by examples of studies in which we have monitored C_2H_2 reduction activity at intervals throughout 12 hr - 12 hr and 16 hr - 8 hr light-dark cycles, determined the C_2H_2/N_2 conversion factor at the midpoint of the light and dark cycles using $^{15}N_2$ fixation, and calculated the daily N input from these data and compared it to the daily N input obtained by N increment.

Nitrogen input determined by the increase in total N is quite straightforward and more reliable. However, as it does not distinguish between the input from N_2 fixation and that from the uptake of exogenous N sources in the aqueous environment, it should not be equated to N_2 fixation unless it is known that there is no exogenous N available. This aspect is illustrated by studies in which <u>Azolla</u> species were grown with and without ^{15}N-labeled ammonium, nitrate or urea, the actual N input from the exogenous N sources and N_2 fixation determined by uptake rates and isotopic dilution, and comparison of the total with that obtained with the N increment method.

Nitrogenase activity determined with isotopic dilution and that determined by the acetylene reduction assay as well as by the N increment procedure are compared, and additional attributes and limitations of the individual approaches considered.

1. Introduction

Rice is a major source of dietary protein for a large segment of the earth's population. Thus, it is a pivotal crop in man's effort to feed himself. Nitrogen is the nutrient most often limiting yield, especially in the newer high yielding rice varieties. The production, availability and cost of commercial N fertilizers is intimately linked to the supply, availability and cost of non-renewable energy sources, especially natural gas. The agricultural significance of the N_2-fixing Azolla-Anabaena azollae symbioses resides primarily in their use as a green manure crop, providing an alternative or supplemental N source for lowland rice. A. pinnata R. Brown has been used traditionally for this purpose in Vietnam as well as in central and southern China (17, 8, 14, 15, 16). During the past decade field studies evaluating this and other species and strains of Azolla have been conducted in a number of geographical locations (28, 29, 30, 36, 37, 38, 35, 32, 33, 34, 10, 11). While these studies have clearly shown the potential of Azolla to provide nitrogen to lowland rice, they have also noted its shortcomings. While Azolla is by no means a panacea, many of these shortcomings may be alleviated through additional research.

When Azolla-Anabaena azollae associations are employed as an N source for rice, either as a fallow season green manure or as a dual crop with or without incorporation during the growing season, the question of whether or not all of the Azolla N is derived from N_2 fixation is of little importance. The important aspect is the total N made available to, and used by, the rice. However, estimates of the total N input from rates of C_2H_2 reduction (2, 3, 4) and most, if not all, determinations of the N input from the increase in Kjeldahl N, or Dumas combustion, are equated to inputs from N_2 fixation alone.

The limitations of the C_2H_2 reduction assay are generally recognized and have been discussed at length by others (9, 31, 13, 27). Nitrogenase activity in vivo is variable and effected by other metabolic processes. Along with them, it is subject to environmental and diurnal effects during the growing period. Thus, estimates of Azolla nitrogen input in the field based on a few point measurements of nitrogenase activity with the C_2H_2 reduction assay are necessarily subject to considerable error. Equating the increase in total N under field conditions to an input from N_2 fixation, with or without point checks of nitrogenase activity with the C_2H_2 reduction assay, ignores the fact that Azolla-Anabaena azollae associations can fix atmospheric N_2 and assimilate exogenous combined N sources simultaneously. During the growth of Azolla in the field some inputs of combined N will necessarily occur from the soil, precipitation, animals and their droppings, dust, weathering and N_2-fixing prokaryotes. More important, however, are the facts that in some regions the irrigation waters may normally contain effluent from waste treatment, that in others there may be sporadic but significant inputs to the irrigation waters from run off of chemical fertilizer N applied to upland crops and that as the Azolla biomass increases there is normal death and decay with some recycling of combined N from the Azolla mat itself.

Subsequent to determining the optimal growth conditions for A. caroliniana, A. filiculoides, A. mexicana and A. pinnata (26), we have conducted a number of comparative physiological studies on these four species under defined laboratory conditions. Preliminary accounts of some of these studies have been presented (22, 21, 12). Here we present results of studies which illustrate: 1) diurnal variation in nitrogenase activity under two photoperiods, as well as the type of agreement between observed doubling times, and those calculated from the % N and the N input derived from integration of the C_2H_2 reduction data and its conversion to N_2 fixation using experimentally determined conversion factors; 2) the inputs from combined N sources, as ammonium, nitrate and urea, and N_2 fixation in the four Azolla species; 3) the type of correlations obtained in comparing inputs of N from the N increment with those obtained from inputs from combined N sources and N_2 fixation using ^{15}N techniques; and 4) the N_2 fixation rates determined using isotopic dilution with nitrogenase activity assayed by C_2H_2 reduction along with the C_2H_2/N_2 ratios. The results are considered with regard to estimates of N input from N_2 fixation by Azolla under field conditions using either the C_2H_2 reduction assay or increase in total N.

2. Materials and Methods

The sources of the Azolla species, growth conditions, determinations of fresh weights, doubling times, and dry weights, and carbon and nitrogen analysis were as described previously (26). Plants were grown on modified N-free IRRI medium buffered at pH 6 with 10 mM MES [2(N-morpholino)ethane sulfonic acid] at the optimal temperature of the individual species at a light intensity of either 400 or 600 μE m^{-2} g^{-1}.

2.1 Acetylene Reduction Assays: The variation in C_2H_2 reduction in the light and dark intervals of two photoperiods was monitored simultaneously with the CO_2 exchange rate. Azolla plants were transferred from culture flasks to small petri dishes which fit the center well of a specially constructed aluminum cylinder measuring 7.5 cm in diameter by 3.5 cm in height and equipped with an inlet and outlet for gas. A clear acrylic lid with an O-ring and compression seal covered the petri dish in the cylinder. A gas phase of 15% C_2H_2 in air was circulated through the cylinder. CO_2 was analyzed with a Beckman Model 315 infra-red gas analyzer. C_2H_2 and C_2H_4 were assayed with a Gow-Mac 750 FID gas chromatograph using column packing and carrier gases described previously (24). The cylinder was checked for leaks prior to introducing sufficient C_2H_2 to provide 15% by volume, and C_2H_2 and a trace amount of methane in the C_2H_2 were used as internal standards. The cylinder was submerged in the water bath where the Azolla was grown and assayed after 15 minutes. All other C_2H_2 reduction assays were conducted in calibrated serum vials equipped with serum caps or crimp caps under 15% C_2H_2 in air and incubated under the conditions used for growth of the Azolla. Except for C_2H_2 reductions conducted with determinations of $^{15}N_2$ fixation, which were for two hours, the incubation period was 15 or 30 minutes.

2.2 $^{15}N_2$ uptake experiments: $^{15}N_2$ (99 atom %) was obtained from Prochem and transferred with a Toepler pump to a storage vessel based on that described in Figure 1 of Burris (5). The desired volume of $^{15}N_2$ was withdrawn with a gas-tight syringe. It was added to evacuated 15 ml serum-stoppered assay vials which had been partially filled with a mixture of $^{14}N_2$, O_2, CO_2 in proportions such that the addition of the $^{15}N_2$ resulted in a slight positive pressure and a gas phase effectively the same as that of the atmosphere, i.e, 80% N_2, 20% O_2 and 0.03% CO_2 by volume. The atom % ^{15}N in the gas phase, which was determined for each vial, was approximately 40%. For determinations at the midpoint of the dark cycles samples were quickly transferred to the vials covered with foil under dim green light. Samples were incubated for two hours under the conditions employed for growth of the cultures. The samples were subjected to semi-micro Kjeldahl digestion with subsequent procedures and analysis of ^{15}N essentially as described in Peters (18).

2.3 Combined N experiments: In studies with combined N as ammonium, nitrate and urea, plants were grown under the conditions optimal for the individual species (26) with a 16 hr - 8 hr light-dark cycle. The indicated concentrations of the combined N sources were added to the N-free medium buffered at pH 6 with 10 mM MES. After 3 or 4 weeks of growth with weekly determinations of growth rates, physiological processes and other parameters, the Azolla were transferred to media containing the same concentrations of ^{15}N-labeled combined N sources. ^{15}N-labeled ammonium chloride, potassium nitrate and urea were obtained from Prochem. Basic aspects of the experimental design are given in the text. Details of the atom % excess of the individual combined N sources and procedures used to determine absorption rates and input from N_2 fixation via isotopic dilution will be presented elsewhere (Ito et al., in preparation). Analytical procedures of Kjeldahl digestion, ammonia distillation, hypobromite oxidation, and analysis of ^{15}N-enrichment were the same as those employed with the $^{15}N_2$ studies.

2.4 N input rates based on weekly N gain or the N increment determinations: Using weekly determinations of doubling times, % N and dry matter content the daily N input was calculated from the average weekly N input using the equation $(\ln N_2 - \ln N_1)/(t_2-t_1) \times (N_2/W_2)$ where N_2 and N_1 are the N content of Azolla at time t_2 and t_1, and W_2 is the fresh weight at t_2.

3. Results and Discussion

The acetylene reduction assay provides a point measurement of nitrogenase activity. It does not provide a measure of N_2 fixation. Acetylene is not a physiological substrate for nitrogenase. In in vivo assays its reduction to non-metabolized ethylene is very different from the reduction of N_2 to ammonia with its subsequent assimilation. Thus, long terms assays should be avoided. Estimates of N input from Azolla using this assay are based on measurements of nitrogenase activity at a given point, or points, in time and space, the use of a conversion factor for the molar relationship between acetylene and nitrogen reduction, and extrapolation through the growth period. Such estimates involve the tacit assumptions that growth, nitrogenase activity and the molar relationship between acetylene and nitrogen reduction remain constant. The fallacy with these assumptions and the questionable significance of such estimates is readily demonstrated.

The effect of environmental conditions on the growth rate and N content of Azolla is well established (17, 3, 1, 33, 34, 26). In contrast to nitrogenase activity, growth rate and N content reflect integrated effects of other metabolic processes and the influence of the environment. Talley et al. followed the diurnal variation in C_2H_2 reduction in A. mexicana and A. filiculoides under field conditions in California on a clear, calm October day (35). A distinct maximum occurred at midday, with the rate of C_2H_2 reduction being about four-fold greater than it was in early morning or late evening. The variation of C_2H_2 reduction rates in A. caroliniana, A. filiculoides, A. mexicana, and A. pinnata during 12 hr light – 12 hr dark and 16 hr light – 8 hr dark cycles in a controlled environment using conditions found optimal for growth of the individual species (26) is shown in Figure 1. The samples used for the light and dark conditions were removed from separate cultures maintained under the same conditions and are considered identical. The growth rates for the individual species under the two photoperiods and under continuous light are presented in Table I. Although the light-dark transition is more abrupt than that which occurs in nature, effects of natural changes, such as light intensity, wind, and temperature, are excluded. Therefore, the variation in rates of C_2H_2 reduction shown in Figure 1 may approximate the lower limit. This variation constitutes the first problem in basing rates of N input on a few rates of C_2H_2 reduction.

The second problem is that of the conversion factor. Since the reduction of N_2 to $2NH_3$ requires 6 electrons, while the reduction of C_2H_2 to C_2H_4 requires 2 electrons, the theoretical conversion factor is $3C_2H_2/N_2$. This relationship is often used in estimating N inputs from C_2H_2 reduction data. However, a number of factors may influence this relationship in vivo. The following are noted: 1) at saturating C_2H_2 all electron flow goes to its reduction and ATP-dependent nitrogenase-catalyzed H_2 reduction is suppressed, whereas at atmospheric N_2, 25% (or more) of the electron flow may go to H_2 production (5). In Azolla some or most of this H_2 may be recycled through an uptake hydrogenase (18; Ito, Tyagi and Peters, unpublished observation). 2) In the isolated nitrogenase, and perhaps in vivo as well, the partitioning of electron flow to N_2 reduction or H_2 production may be dependent on ATP levels and the ADP/ATP ratios (Mortenson, personal communication). 3) C_2H_2 may alter electron flux through nitrogenase (7; Burgess, personal communication) and; 4) acetylene is a reactive molecule and its effect on other enzymes and cellular processes is still being documented. Determinations for various organisms under field conditions have resulted in a range of values for the C_2H_2/N_2 ratio from 1 to 20 (27). While there are a variety of factors in addition to those noted here which undoubtedly effects some of these determinations, there is no compelling reason to exclude the possibility that this ratio may vary with the organism's growth rate and overall metabolic activity under different environmental conditions, and as a function of plant or culture age.

In conjunction with the data presented in Figure 1, rates of $^{15}N_2$ fixation and C_2H_2/N_2 ratios were determined for each species at the midpoints of the light and dark intervals of both photoperiods. While separate determinations of the C_2H_2/N_2 ratio ranged from 2.15 for $\underline{A}$. $\underline{caroliniana}$ during the light of the 12 hr cycle to 5.26 for $\underline{A}$. $\underline{pinnata}$ during the dark of the 12 hr cycle, the overall average was 3.84 ± 0.89. The means for the four species at individual midpoints were as follows: 12 hr light, 3.14 ± 0.97; 12 hr dark, 4.04 ± 1.32; 16 hr light, 4.10 ± 0.38 and 8 hr dark, 4.10 ± 0.52. When the four species were grown under continuous light the average value for the C_2H_2/N_2 ratio was 2.40 ± 0.48. Thus, there is variation in not only the rate of C_2H_2 reduction but in the determined relationship between C_2H_2 reduced and N_2 fixed in the four $\underline{Azolla}$ species under controlled environment conditions which are optimal for their growth.

In estimating N input it is important to remember that both partners in the $\underline{Azolla-Anabaena}$ $\underline{azollae}$ symbioses are photosynthetic organisms, and that nitrogenase activity at night is not only appreciably less than that during the day but dependent upon photosynthate from prior photosynthesis (24, 25, 26, 19, 20) (also see Figure 1). Using the rates of $^{15}N_2$ fixation at the midpoints of the light and dark portions of the two photoperiods we estimated the contributions from the light and dark fixation during a 24 hr period for the four species (12, 21). Under the 12 hr-12 hr cycle the average for the four species was 81.0 ± 6.2% during the light period, and under the 16 hr - 8 hr cycle, 82.3 ± 1.9% during the light period. Using this approach it was acknowledged that the contribution during the light of the 12 hr period might be underestimated since C_2H_2 reduction was found to drop after 7 hr of darkness. Integration of the area under the curves for C_2H_2 reduction in $\underline{A}$. $\underline{caroliniana}$ (Figure 1) indicated 85% of the daily N input would occur during a 12 hr light period and 90% during a 16 hr light period. There was a good correlation between the daily N_2 fixed calculated from rates of $^{15}N_2$ fixation at the midpoints of the light-dark cycles and the Kjeldahl N content of the four individual species grown under the two photoperiods, with $r^2 = 0.875$ for the 16 hr - 8 hr and $r^2 = 0.673$ for the 12 hr - 12 hr cycles.

A number of factors come into play in attempts to relate rates of physiological processes such as photosynthesis, respiration and nitrogen fixation in $\underline{Azolla}$ to observed growth rates and % C and % N content. In addition to the inherent biological variability of plants within a population, measurements of the physiological processes are normally short term assays, whereas growth rates, % C and % N represent their net result integrated over a longer time frame. In a previous study we presented the C_2H_2 reduction data of $\underline{A}$. $\underline{caroliniana}$ in Figure 1 along with the rates of net photosynthesis and dark respiration over the same periods (23). Areas under the curves were integrated. Light and dark C_2H_2 reduction was converted to N_2 fixed using the appropriate conversion factors and added to approximate daily N input. Dark respiration was subtracted from apparent photosynthesis and daily N input estimated. By knowing the % C and % N of the tissue it was possible to use the calculated daily C and N inputs to calculate a doubling time and to compare this with the observed doubling time. For the 12 hr - 12 hr cycle the observed doubling time was 2.7 days (Table I), that calculated for the C input was 2.8 days and that from the N input 2.5 days. For the 16 hr - 8 hr the observed doubling time was 2.2 days (Table I), while those calculated from carbon and nitrogen input were 2.0 and 3.5, respectively. Intuitively one might have expected a better agreement for the N calculations for the 16 hr - 8 hr, where the C_2H_2/N_2 ratios in the light and dark were 4.09 and 4.40, respectively, than in the 12 hr - 12 hr where these values were 2.15 and 3.28, respectively.

We would summarize the information presented thus far in the following manner. Based on controlled environment studies we would suggest that determining rates of C_2H_2 reduction, $^{15}N_2$ fixation and their molar relationship under field conditions with subsequent extrapolation through the growing season to estimate N input from N_2 fixation may well grossly over or under estimate the actual N input. Our results suggest that the actual C_2H_2/N_2 ratio may vary by at least a factor of two under controlled environmental conditions. They also imply that while the determination of a single conversion factor for use with multiple determinations of C_2H_2 reduction during a growing season might appear to add credibility to the estimates, such estimates are probably no more accurate than they would be if one assumed a conversion factor of about 4 at the onset.

Determination of the N input from N_2 fixation based on the increase in total N generally provides a more realistic and agronomically acceptable value than extrapolation from C_2H_2 reduction. However, this procedure is not without its limitations. Estimates based on the increase in biomass and N content of an Azolla mat at full cover over that used as an inoculum is affected by a number of factors, some of which are noted here. First, the increase in total N, determined by Kjeldahl digestion or Dumas combustion, does not measure N_2 fixation. Estimates of N input using these approaches reflect the net gain which is a composite of the N input from N_2 fixation, usable combined N sources in the paddy water, the extent of recycling of these N sources and their loss to the paddy water during normal death and decay of Azolla, losses through volatilization, and so on. Moreover, as such estimates are often extrapolated from subsamples of the biomass and N content per unit area, there is also the potential for error due to spatial variation. Only the potential impact of combined N sources is addressed here.

The relative contributions of N_2 fixation and combined N sources to the total daily N input in Azolla has been assessed. In an initial study ^{15}N-labeled ammonium, nitrate and urea were used to determine absorption rates and daily input from the combined N sources, and isotopic dilution in the presence and absence of the same concentrations of the unlabeled combined N sources was used to determine the inputs from N_2 fixation in A. caroliniana (22). Subsequently similar studies were conducted with A. caroliniana, A. filiculoides, A. mexicana and A. pinnata. Table II, a greatly condensed summary of these studies reported elsewhere (21) shows the effect of several concentrations of ammonium, nitrate and urea on growth rates, % C, % N and the estimated percentage of the daily N inputs attributable to the combined N sources, as well as the relative rates of C_2H_2 reduction. Detailed results showing actual rates obtained from photosynthesis, dark respiration, acetylene reduction, absorption of the combined N sources and N_2 fixation will be presented elsewhere (Ito et al., unpublished). Two separate points merit comment in regard to these studies. First, it is imperative that Azolla species used in such studies are devoid of all epiphytic cyanobacterial or algal contaminants. Second, although the Azolla species is not stated, Watanabe et al. have reported somewhat different effects from combined N sources as ammonia or nitrate with a 50% decrease in C_2H_2 reduction activity after only four days on 1 mM nitrate or ammonium and N_2 fixation accounting for 46% of the total N assimilated in this interval (38).

There was a relatively good correlation between the N input determined on the basis of weekly N gain (see Methods) in the Azolla species grown on unlabeled combined N sources during the two weeks prior to the use of ^{15}N-labeled combined N sources and that calculated on the basis of the sum of the rates obtained for uptake of the combined N sources and N_2 fixation (Figure 2).

While there is a relatively good agreement between the relative rates of C_2H_2 reduction and the calculated inputs from combined N sources in Table II, and by inference the inputs from N_2 fixation, this is somewhat fortuitous. The relative rates of C_2H_2 reduction in this table are based on the means of weekly determinations throughout the three to four week study period, whereas the calculated inputs from the combined N sources were based on studies at the end of this period. After 3 to 4 weeks of growth, with weekly transfers, on the indicated concentrations of the combined N sources, the plants were transferred to media containing the appropriate concentration of ^{15}N-labeled combined N sources for 2 days. After removing material for ^{15}N analysis and the determination of absorption rates, half of the remaining material was transferred to medium containing the appropriate concentration of the unlabeled combined N source and the other half to N-free medium for two more days. The rate of N_2 fixation was calculated from dilution of the ^{15}N-label in the plant material. In the absence of a combined N source the dilution was attributed to N_2 fixation alone. In the presence of the unlabeled combined N source the rate of N_2 fixation was calculated after correcting for the dilution by the unlabeled N source using the absorption rate determined with the ^{15}N-labeled compounds. The daily N input attributable to the combined N source was determined by the ratio A/(A + 2F) where A is the absorption rate of the combined N source and F the rate of N_2 fixation. In contrast to the seemingly good agreement between the input from the combined N source and the relative rate of C_2H_2 reduction in Table II, the correlation between the rates of C_2H_2 reduction and the rates of N_2 fixation determined in the presence and absence of the combined N sources during the final few days of the study was variable (Figures 3 to 5). Upon transfer to media without a combined N source there was some recovery of nitrogenase activity measured by either C_2H_2 reduction or isotope dilution. Admittedly a number of factors, including experimental error, may contribute to the varying degrees of correlation in Figures 3 to 5. However, we would emphasize that these C_2H_2 reduction assays are based on a 15 minute assay, i.e., one time point, whereas the rate of N_2 fixation is based on dilution of label over a two day period. Thus, these data may simply reflect the point made earlier, namely, that nitrogenase activity is variable and a point measurement of this activity with the C_2H_2 reduction assay should not be expected to correlate well with an approach which provides the average rate over a two day period. We currently attribute minimal significance to the differences in correlation with the three N sources. As noted, in the legend of Figure 3,

Figures 3 to 5 are composites of N_2 fixation and C_2H_2 reduction in plants transferred to media with or without the indicated N source. The average ratios of C_2H_2 reduced/N_2 fixed in the individual treatments were as follows. In the ammonia experiment the mean value for plants transferred to N-free media was 5.55 ± 0.73, while that of plants transferred to media with the appropriate concentration of unlabeled ammonia was 5.95 ± 1.77. For the studies employing nitrate these values were 7.14 ± 2.99 and 5.66 ± 2.22 and for the studies employing urea they were 5.83 ± 1.04 and 4.36 ± 1.58.

The vigorous growth of the four Azolla species on media containing combined N as ammonia, nitrate or urea and the use of ^{15}N techniques to demonstrate the ability of all four species to simultaneously assimilate the exogenous N sources and fix atmospheric nitrogen are evidence of the need to refrain from equating increases in total N in the field to N_2 fixation. In general it is sufficient to note the estimate of the total N and, if nitrogenase activity has been monitored with the C_2H_2 reduction assay, to note the levels of this activity and the fact that such activity is indicative of N_2 fixation contributing to the total N of the biomass.

Acknowledgement The authors' research was supported in part by NSF/ASRA Grant PFR 77-27269. The technical assistance of Mr. Robert E. Toia, Jr. is gratefully noted, as is the clerical assistance of D. Patten.

References

1 Ashton P J 1974 The effect of some environmental factors on the growth of Azolla filiculoides Lam. In: E.M.V. Zinderen-Bakker (ed.) The Orange River, Progress Report. Bloemfontein, S. Africa, pp. 123-138.

2 Becking J H 1976 Contribution of plant-algal associations. In: W.E. Newton and C.J. Nyman (eds.). Proceedings of the 1st International Symposium on Nitrogen Fixation, Vol. 2. Washington State University Press, Pullman, Washington, pp. 556-580.

3 Becking J H 1979 Environmental requirements of Azolla for use in tropical rice production. In: Nitrogen and Rice. IRRI, Los Banos, Laguna, Philippines, pp. 345-373.

4 Brotonegoro S and Abdulkadir S 1976 Growth and nitrogen-fixing activity of Azolla pinnata. Ann. Bogor. 6: 69-77.

5 Burris R H 1972 Nitrogen fixation - Assay methods and techniques. Meth. Enz. 24: 415-431.

6 Burris R H 1974 Methodology. In: A. Quispel (ed.) The Biology of Nitrogen Fixation. North-Holland Publ. Co., Amsterdam, pp. 9-33.

7 Burris R H, Arp D J, Hageman R V, Houchins J P, Sweet W J and Tsu M Y 1981 Mechanism of nitrogenase action. In: A.H. Gibson and W.E. Newton (eds.) Current Perspectives in Nitrogen Fixation. Australian Academy of Science, Canberra, pp. 56-66.

8 Dao T T and Tran T Q 1979 Use of Azolla in rice production in Vietnam. In: Nitrogen and Rice. IRRI, Los Banos, Laguna, Philippines, pp. 395-405.

9 Hardy R W F, Burns R C and Holsten R D 1973 Applications of the acetylene-ethylene assay for measurement of nitrogen fixation Soil Biol. Biochem. 5: 47-81.

10 International Rice Research Institute 1980 Report on the first trials of Azolla use to rice, INSFFER 1979. IRRI, Los Banos, Laguna, Philippines.

11 International Rice Research Institute 1981 Report on the second trials of Azolla use to rice, INSFFER 1980. IRRI, Los Banos, Laguna, Philippines.

12 Ito O, Toia R E Jr., Poole R E, Crist D K, Evans W R, Mayne B C and Peters G A 1980 Physiological studies on N_2-fixing Azolla species grown under three photoperiods. Plant Physiol. 65: S-109.

13 Knowles R 1981 The measurement of nitrogen fixation. In: A.H. Gibson and W.E. Newton (eds.) Current Perspectives in Nitrogen Fixation. Australian Academy of Science, Canberra, pp. 327-333.

14 Liu C C 1979 Use of Azolla in rice production in China. In: Nitrogen and Rice. IRRI, Los Banos, Laguna, Philippines, pp. 376-394.

15 Lumpkin T A and Plucknett D L 1980a Azolla, a low-cost aquatic green manure for agricultural crops. Paper on Innovative Biological Technologies for Lesser Developed Countries for the Committee on Foreign Affairs, U.S. House of Representatives.

16 Lumpkin T A and Plucknett D L 1980b Azolla: Botany, physiology, and use as a green manure. Econ. Bot. 34: 111-153.

17 Moore A W 1969 Azolla: Biology and agronomic significance. Bot. Rev. 35: 17-34.

18 Peters G A 1977 The Azolla-Anabaena azollae symbiosis. In: A. Hollaender (ed.) Genetic Engineering for Nitrogen Fixation. Plenum Press, New York, pp. 231-258.

19 Petes G A and Calvert H E 1982a The Azolla-Anabaena symbioses. In: N.S. Subba Rao (ed.) Advances in Agricultural Microbiology. Oxford & IBH Publ. Co., New Delhi, pp. 191-218.

20 Peters G A and Calvert H E 1982b The Azolla-Anabaena azollae symbioses. In: L. Goff (ed.) Algal Symbiosis: A Continuum of Interaction Strategies. Cambridge University Press, New York.

21 Peters G A, Calvert H E, Kaplan D, Ito O and Toia R E Jr. 1982 The Azolla-Anabaena symbiosis: Morphology, physiology and use. Israel J. Bot. 30.

22 Peters G A, Ito O, Tyagi V V S and Kaplan D 1981a Physiological studies on N_2-fixing Azolla. In: A. Hollaender (ed.) Genetic Engineering of Symbiotic Nitrogen Fixation and Conservation of Fixed Nitrogen. Plenum Press, New York, pp. 343-362.

23 Peters G A, Ito O, Tyagi V V S, Mayne B C, Kaplan D and Calvert H E 1981b Photosynthesis and N_2 fixation in the Azolla-Anabaena symbiosis. In: A.H. Gibson and W.E. Newton (eds.) Current Perspectives in Nitrogen Fixation. Australian Academy of Science, Canberra, pp. 121-124.

24 Peters G A and Mayne B C 1974 The Azolla, Anabaena azollae relationship. I. Initial characterization of the association. Plant Physiol. 53: 813-819.

25 Peters G A, Mayne B C, Ray T B and Toia R E Jr. 1979 Physiology and biochemistry of the Azolla-Anabaena symbiosis. In: Nitrogen and Rice. IRRI, Los Banos, Laguna, Philippines, pp. 325-344.

26 Peters G A, Toia R E Jr., Evans W R, Crist D K, Mayne B C and Poole R E 1980 Characterization and comparisons of five N_2-fixing Azolla-Anabaena associations. I. Optimization of growth conditions for biomass increase and N content in a controlled environment. Plant, Cell & Environ. 3: 261-269.

27 Silvester W B 1981 Measurement of nitrogen fixation. In: A.H. Gibson and W.E. Newton (eds.) Current Perspectives in Nitrogen Fixation. Australian Academy of Science, Canberra, pp. 334-337.

28 Singh P K 1977 Multiplication and utilization of fern "Azolla" containing nitrogen-fixing algal symbiont as green manure in rice cultivation. Il Riso 26: 125-137.

29 Singh P K 1979a Use of Azolla in rice production in India. In: Nitrogen and Rice. IRRI, Los Banos, Laguna, Philippines, pp. 407-418.

30 Singh P K 1979 Symbiotic algal N_2-fixation and crop productivity. In: C.P. Malik (ed.) Ann. Rev. Plant Sci. 1: 37-65.

31 Stewart W D P, Fitzgerald G P and Burris R H 1967 In situ studies on N_2 fixation using the acetylene reduction technique. Proc. Nat'l. Acad. Sci. U.S. 58: 2071-2078.

32 Talley S N, Lim E and Rains D W 1981 Application of Azolla in crop production. In: J.M. Lyons, R.C. Valentine, D.A. Phillips, D.W. Rains, and R.C. Huffaker (eds.) Genetic Engineering of Symbiotic Nitrogen Fixation and Conservation of Fixed Nitrogen. Plenum Press, New York, pp. 363-384.

33 Talley S N and Rains D W 1980a Azolla filiculoides Lam. as a fallow-season green manure for rice in temperate climate. Agron. J. 72: 11-18.

34 Talley S N and Rains D W 1980b Azolla as a nitrogen source for temperate rice. In: W.E. Newton and W.H. Orme-Johnson (eds.) Nitrogen Fixation, Vol. II, Symbiotic Associations and Cyanobacteria. University Park Press, Baltimore, pp. 311-320.

35 Talley S N, Talley B J and Rains D W 1977 Nitrogen fixation by Azolla in rice fields. In: A. Hollaender (ed.) Genetic Engineering for Nitrogen Fixation. Plenum Press, New York, pp. 259-281.

36 Watanabe I 1978 Azolla and its use in lowland rice culture. Soil Microbes 20: 1-10.

37 Watanabe I, Espinas C R, Berja N S and Alimagno B V 1977 Utilization of the Azolla-Anabaena complex as a nitrogen fertilizer for rice. IRRI Res. Paper Ser. 11: 1-15.

38 Watanabe I, Ito O and Espinas C R 1981 Growth of Azolla with rice and its effect on rice yield. Int. Rice Res. Newsletter 6: 23.

Table 1. Doubling time (in days) of four Azolla species grown under their optimal conditions, where temperatures were 25° for A. caroliniana and A. filiculoides, and 30° for A. mexicana and A. pinnata.

Light-dark (hrs)	A. caroliniana	A. filiculoides	A. mexicana	A. pinnata
12-12	2.74±0.21	2.60±0.20	1.86±0.09	2.23±0.40
16-8	2.21±0.21	2.23±0.25	1.58±0.03	1.81±0.06
24-0	1.64±0.04	1.71±0.09	1.52±0.03	1.65±0.06

Table 2a-c. Effect of Ammonium (a), Nitrate (b), and Urea (c) on Growth, % Dry Matter, %C, %N, the estimated % of the Daily N Input from the Combined N Source, and the Relative Rate of C_2H_2 reduction.

(a) mM NH_4^+ in growth medium

	A. caroliniana			A. filiculoides		
	0	2.5	5.0	0	2.5	5.0
Doubling Time (days)	1.85±0.03	1.78±0.05	1.84±0.07	1.75±0.04	1.84±0.06	1.75±0.04
% Dry matter	5.8 ±0.4	5.5 ±0.2	5.7 ±0.3	5.4 ±0.2	5.4 ±0.2	5.7 ±0.2
%C	43.2 ±0.1	43.2 ±0.2	43.5 ±0.5	43.1 ±0.1	43.3 ±0.8	43.6 ±0.4
%N	4.6 ±0.3	5.5 ±0.1	5.6 ±0.3	5.6 ±0.4	6.4 ±0.1	6.5 ±0.2
% of daily N input from NH_4^+	0	35	52	0	61	63
Relative C_2H_2 reduction	100	61	57	100	58	49

(b) mM NO_3^- in growth medium

	A. caroliniana			A. filiculoides		
	0	10	25	0	10	25
Doubling Time (days)	2.16±0.14	2.08±0.17	2.18±0.12	1.95±0.11	2.01±0.06	2.00±0.05
% Dry matter	5.6 ±0.7	5.7 ±0.8	6.3 ±0.6	4.8 ±0.4	5.3 ±0.6	5.7 ±6.0
%C	42.8 ±0.1	42.7 ±0.1	42.3 ±0.1	42.0 ±0.4	42.0 ±0.3	41.3 ±0.4
%N	5.1 ±0.7	5.3 ±0.5	4.7 ±0.7	6.0 ±0.8	6.1 ±0.7	6.0 ±0.6
% of daily N input from NO_3^-	0	25	63	0	21	31
Relative C_2H_2 reduction	100	66	38	100	61	41

(c) mM Urea in growth medium

	A. caroliniana			A. filiculoides		
	0	1.25	12.5	0	1.25	12.5
Doubling Time (days)	2.04±0.12	1.92±0.46	1.89±0.08	1.92±0.07	1.93±0.09	1.83±0.03
% Dry matter	5.3 ±0.5	5.6 ±0.8	6.1 ±1.1	4.9 ±0.7	4.9 ±0.2	5.4 ±0.2
%C	41.3 ±1.1	42.8 ±0.9	43.4 ±0.8	41.2 ±0.5	42.6 ±0.5	42.6 ±0.2
%N	5.5 ±0.2	5.7 ±0.6	6.2 ±0.7	5.8 ±0.6	6.6 ±0.3	7.0 ±0.4
% daily N input from Urea	0	20	41	0	22	51
Relative C_2H_2 reduction	100	94	52	100	67	43

Table 2a–c. Continued

(a) mM NH_4^+ in growth medium

	A. mexicana			A. pinnata		
	0	2.5	5.0	0	2.5	5.0
Doubling Time (days)	1.97±0.32	1.86±0.12	1.91±0.13	1.97±0.32	1.84±0.13	1.85±0.20
% Dry Matter	5.4 ±0.8	5.6 ±0.8	5.9 ±1.0	5.9 ±0.6	6.0 ±0.3	5.8 ±0.1
%C	42.3 ±0.8	43.8 ±0.9	44.6 ±0.6	42.3 ±0.5	44.1 ±0.6	44.2 ±0.1
%N	6.1 ±0.2	6.7 ±0.4	6.8 ±0.4	4.6 ±0.2	5.4 ±0.4	6.0 ±0.5
% of daily N input from NH_4^+	0	37	45	0	41	47
Relative C_2H_2 reduction	100	67	62	100	54	49

(b) mM NO_3^- in growth medium

	A. mexicana			A. pinnata		
	0	10	25	0	10	25
Doubling Time (days)	1.43±0.08	1.93±0.08	2.07±0.07	1.91±0.04	1.98±0.08	2.28±0.15
% Dry matter	4.8 ±0.4	5.1 ±0.5	5.2 ±0.4	4.9 ±0.8	5.2 ±0.6	5.2 ±0.7
%C	43.0 ±0.9	42.3 ±0.7	40.9 ±1.0	43.4 ±0.2	43.3 ±0.6	43.0 ±0.7
%N	6.6 ±0.5	6.8 ±0.4	6.5 ±0.5	5.6 ±0.9	5.6 ±0.9	6.1 ±0.4
% of daily N input from NO_3^-	0	9	17	0	16	24
Relative C_2H_2 reduction	100	100	80	100	77	58

(c) mM Urea in growth medium

	A. mexicana			A. pinnata		
	0	1.25	12.5	0	1.25	12.5
Doubling Time (days)	1.81±0.06	1.82±1.12	1.88±1.08	1.87±0.03	1.81±0.04	1.84±0.06
% Dry matter	4.6 ±0.4	4.7 ±0.6	5.1 ±0.1	4.8 ±0.3	5.2 ±0.7	5.4 ±0.7
%C	42.7 ±1.5	42.6 ±1.3	43.2 ±0.9	42.5 ±1.2	42.8 ±1.3	43.2 ±1.1
%N	7.0 ±0.2	7.3 ±0.4	8.1 ±0.6	5.7 ±0.4	6.0 ±0.5	7.0 ±02
% daily N input from Urea	0	23	50	0	25	49
Relative C_2H_2 reduction	100	80	42	100	61	50

[1] The plants were grown under optimal conditions with the indicated concentration of the combined N source for 3 or 4 weeks with weekly transfers. Except for the estimate of the daily N input, the plants were sampled weekly, in duplicate, and the data reflects the average value of all determinations. Estimates of the daily N input are based on duplicate determinations, using isotope dilution techniques at the end of the experimental period. The relative C_2H_2 reduction is expressed as a percentage of the control grown in the absence of combined N, simply using the mean values from the total study period.

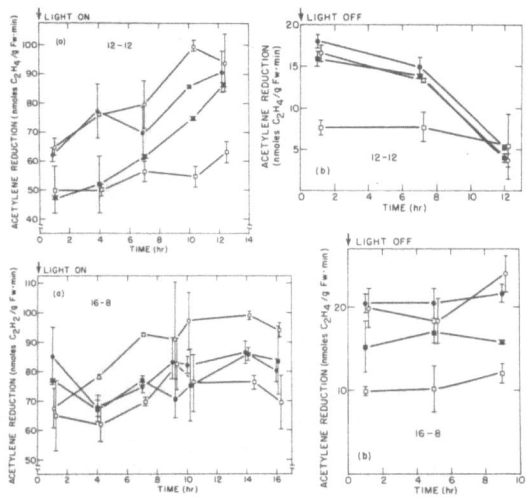

Figure 1 (a and b). Acetylene reduction (nitrogenase activity) in four <u>Azolla</u> species during the light (a) and dark (b) intervals of two photoperiods. <u>A</u>. <u>caroliniana</u>, <u>A</u>. <u>filiculoides</u>, <u>A</u>. <u>mexicana</u> and <u>A</u>. <u>pinnata</u>.

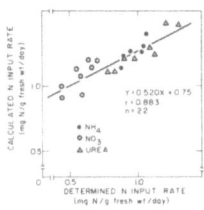

Figure 2. The correlation between nitrogen input rates calculated from the weekly N gain and those determined with ^{15}N studies. The plotted data is a composite of the information from the four <u>Azolla</u> species. The calculated input rates are based on the N gain in each <u>Azolla</u> species on two concentrations of each combined N source (2.5 and 5.0 mM ammonium, 5.0 and 10.0 mM nitrate and 1.25 and 12.5 mM urea) at the end of the first, second and third weeks of growth. The plotted values are based on the average ΔN at the end of the second and third weeks. The determined N input rates are based on the absorption rates of ^{15}N-labeled ammonia, nitrate and urea at the noted concentrations, plus rates of N_2 fixation, obtained from isotope dilution measurements. Data for <u>A</u>. <u>filiculoides</u> on ammonium were not plotted, thus n=22 rather than 24.

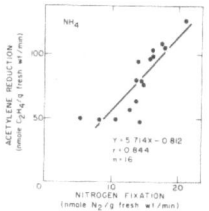

Figure 3.   The correlation between rates of acetylene reduction
and the rates of N_2 fixation calculated from isotopic dilution for
the four <u>Azolla</u> species grown on 2.5 and 5.0 mM ammonia.   After
three weeks of growth with weekly transfers, the <u>Azolla</u> species
were transferred to medium containing the same concentrations of
[15]N-labeled ammonium for two days.   Following removal of plant
material for determination of absorption rates, half of the material
from each culture vessel was transferred to media without combined
N and the other half to media with the appropriate concentration
of unlabeled ammonium for two more days.   The plotted data includes
the rate of N_2 fixation and acetylene reduction in each treatment,
hence n=16.

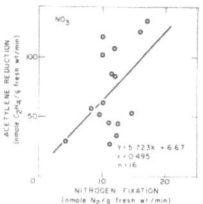

Figure 4.   The correlation between rates of acetylene reduction
and the rates of N_2 fixation calculated from isotopic dilution for
the four <u>Azolla</u> species grown on 5.0 and 10 mM nitrate.   Procedures
were as described in Figure 3 except that transfer to media with
[15]N-labeled nitrate was after the fourth week.

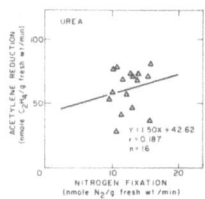

Figure 5.   The correlation between rates of acetylene reduction
and the rates of N_2 fixation calculated from isotopic dilution for
the four <u>Azolla</u> species grown on 1.25 and 12.5 mM urea.   Plants
were transferred to media with ^{15}N-labeled urea after the fourth
week.   Other procedures are as described in Figure 3.

4. Recent advances on Azolla research

Liu Chung-Chu
Fujian Academy of Agricultural Sciences
Fuzhou, Fujian, China

Key words   Alga-free Azolla   Azolla-nitrogen fixation Nitrogen
excreting   Potassium enriching   Recombination Spore propagation

Summary

Our recent research indicate that Azolla cannot only fix nitrogen
from the atmosphere, but also can absorb traces of potassium
(1-5 ppm K_2O) from water. So Azolla may be a potential potassium
source for rice production. Azolla can also excrete 3-4% of the
total amount N fixed to the exterior solution. This would open
a way to utilize Azolla.
Double-narrow-rows method, an improved cultivating technique,
led to greatly enhanced biomass production of Azolla. Azolla
caroliniana with its multi-tolerant capacities provides a promising
strain for growing in hot regions.
Preliminary success in the recombination of algae-free Azolla
and Anabaena azollae suggest that it can be used for Azolla breeding
and studying the relationship between the fern and micro-symbiont.

1. Introduction

China has a long history of using Azolla in rice production. Recently
the use of and research on Azolla has developed quickly. However
there are still some obstacles hindering expansion of the use of
Azolla especially survival and propagation in the summer season.
In recent years we have been dealing with these problems.

2. Nitrogen Fixation

The N_2-fixing capacity of Azolla is influenced by light, temperature,
mineral nutrients, and the nature of the Azolla strain[5,6,10]. Generally, with the changes of light intensity and water temperature,
the N_2-fixing activity of various Azolla shows a periodical variation
of high and low activity within 24 hours. The high normally appears
at about 14:00 hours and the low at zero hour each day. But even
without light (at night) Azolla still maintains some N_2-fixing activity apparently due to the mobilization of resource substances to
produce energy. The amount of N_2 fixed at night is roughly one-half
of that during daylight. Some strains, like A. caroliniana, fix
N_2 at night during autumn up to 42% of that fixed during the day.
Shipin Green (A. imbricata a local strain, evergreen) may account
for 43% of that in the daytime of winter (Table 1). From this we
can see that the amount of N fixed by Azolla at night is significant.
Whether light or temperature is the controlling factor depends
upon the strain of Azolla and the season (Table 2). For A.
caroliniana light is not the main factor regardless of the season.

Since it is both light resistant and shade tolerant it has a wide range of adaptations. However, in autumn and winter temperature becomes the predominant factor. Nitrogen fixation in A. imbricata is very susceptible to light and temperature, with a narrow scope of adaptation, whereas Azolla pinnata is somewhere between the two particularly with regard to high temperature. In all strains the highest N_2-fixing activity and total amount of N_2 fixed is always in the spring or autumn, the optimum light intensity is 40,000 to 60,000 Klux and the optimum temperature, $20°C$ to $30°C$.

The nutrient levels in the exterior solution have a notable influence on N_2-fixation in Azolla . The order of interference with N_2-fixation is as follows: Nitrate-N < ammonium-N < nitrate-N plus ammonium-N < urea-N. When nitrate and ammonium are both applied the effect is additive. Therefore, in the paddy field attention should be paid not only to the amount of fertilizer to be applied but also the kind. Obviously, urea is not suitable and ammonium sulfate is preferable. In the paddy field where Azolla is cultivated, deep placement of N fertilizer would be most desirable.

3. Nitrogen Excretion

Our experiments indicate that during growth of Azolla a portion of the N fixed is excreted–about 3–4% of the total N fixed, and depending upon the strain. The N excreted by Azolla is used by the paddy rice. If it is possible to improve the excretory capacity of Azolla, and to prolong the period of intercropping of Azolla and paddy rice by means of double-narrow-rows, a new ecological system could be established.

4. Potassium uptake by Azolla

Besides the capacity for N_2-fixation, Azolla also possesses rather strong capacity for absorbing K from water. The K content of Azolla was compared to that in the exterior solution. When the latter ranged from 0.1-1-500 ppm, the K concentration in Azolla increased exponentially. Under normal growing conditions it is within 0.2–0.3%. Azolla can rapidly recover and make use of the very low concentration of K. The greatest percentage of K removed from solution occurred when the solution contained 0.85 ppm K_2O, which may be regarded as the physiological critical point of K requirement for Azolla. This ability is remarkable in contrast to rice which absorbs little K when the concentration is low. The peak of K recovery for the rice from the exterior solution is at about 8 ppm, or its critical value of physiological requirement of K is 10 times that of Azolla. If the concentration of exterior solution is less than 1 ppm, no K is recovered[9].

The K concentration in the irrigation water and rainfall is around 1–5 ppm, which is below the level which rice can absorb and utilize readily; however Azolla can rapidly pick up and concentrate K at these trace concentrations. Once incorporated into soil Azolla will decompose and release the K for the use of rice. These results show that Azolla may be a potential source of potassium for rice.

5. Improved cultivation leads to enhancement of output

Cultivating Azolla in the paddy field mainly consists of two steps: one is to make use of fallow or empty fields 1 to 2 months before transplanting for growing Azolla which is later used as a basal manure; then to intercrop with rice and use as a top dressing. The former requires arable land, and the later results in yields normally from 15-22.5 tons/ha. In recent years, we have studied and adopted a double-narrow-row method of rice transplanting to incorporate with cultivation of Azolla. Time for Azolla growth is extended and biomass increased as compared with normal rice spacing, while grain yield of rice is not affected. The biomass of Azolla is normally 40-55 tons/ha/crop .

The double-narrow-row is arranged as follows: narrow row spacing for rice is 13.3 cm and the in-row spacing for conventional strains is 6.7-8.3 cm and for hybrid rice, 8.2-10 cm. The wide row spacing to permit Azolla growth in the case of conventional rice varieties is 40-53.3 cm and hybrid rice 46.7-60 cm. In 1980, the Soil and Fertilizer Research Institute of the Hunan Academy of Agricultural Sciences adopted this method to grow Azolla in single medium hybrid rice. The three test groups obtained rice yields of 7.48; 7.65; and 7.61 tons/ha, and an Azolla biomass of 84.6, 75.2, and 59.7 tons/ha respectively. Watanabe (1980) reported that using this method for two years Azolla has been harvested four to six times for each crop of rice. Azolla so obtained is equivalent to 70 Kg N/ha and the grain yield to that achieved by the application of 70-100 Kg N/ha as chemical fertilizer.

6. Screening and breeding new strains

6.1 Screening new strains. Azolla has been cultivated as a green manure for many years. Breeding new strains of Azolla has not yet been attacked. Only A. imbricata and A. filiculoides are applied in agriculture so far in China. Since these two species are not so heat and pest tolerant, many difficulties have hindered the wide cultivation of Azolla in summer season. In order to overcome this obstacle since 1978 we have worked on screening for Azolla strains with multi-tolerant capacity. Laboratory and field experiments showed that A. caroliniana may be a promising species to employ.

6.2 Strong tolerance to high temperature and light intensity. Azolla caroliniana shows higher nitrogenase activity under high temperature and strong light intensity even during hot summer nights compared with A. filiculoides and A. pinnata. A striking feature is that a decline of nitrogenase activity resulting from high temperature and strong light will rapidly rise again in the morning of the following day.

6.2 Outstanding shade and diseases resistance. A. caroliniana has a maximum growth rate when light intensity is 20,000 Klux or more. A similar growth rate is likely to remain when light decreases to 6,000 Klux. Under a daily average temperature of 30°C and high humidity it can still tolerate weak illumination (3,000 Klux) and survive, while A. imbricata Yuling strain and others drastically decline in yield under 6,000 Klux and quickly die off should the light intensity decrease further (Table 3).

Inoculation with mildew disease showed that the disease-resistance of various strains of Azolla is closely related to their shade tolerance. The disease did not occur in A. caroliniana even under a temperature of 20°C and light intensity of 2,000 Klux for 10 days while Yuling strain was going to perish under these conditions. This point has been verified in practice at Lou Shia and Yinshi experimental sites. We found that in spite of the fact that during the later stage the full-grown rice plants sheltered the surface, when water was drained and the field covered at row intervals like a green carpet, no mildew disease was detected. After harvesting, the biomass of Azolla was estimated at 6.5 tons/ha.

6.3 Resistance to Azolla snail and pests. Azolla snail (mostly Radix swinhoer H. Adams) is a major pest of Azolla. Compared with 10 strains (or species), A. caroliniana showed more tolerance to snails. Experiments also indicated that the population of thread worm grew slower on A. caroliniana and yields suffer less from the pest (Table 4).

6.4 Steady annual yield. Statistics based on the tests in plots, ponds, and pots with 3-4 replications showed that no clear margin was observed between the growth rates of A. caroliniana and A. filiculoides in spring, autumn, and winter, but in summer A. caroliniana was much superior (Table 5).

6.5 The application of spore propagation to Azolla production. The difficulties of surviving winter in the north and surviving summer in the south are the main obstacles to spreading Azolla on a large scale. Recently, a new method by means of spore propagation has been developed in Guangdong[2][3], Zhejiang[11], Fujian[1], and other provinal Academies for expediting the extension of Azolla. Although no systematic knowledge about the inducing conditions for spore formation has been acquired, some notable progress has been made relevant to the sprouting conditions of spores, the growth of seedlings, and the methods of gathering the spores. To date, spores have sprouted within 7-10 days and grown into seedlings in 25-31 days.

6.6 Recombination of algae-free Azolla and Anabaena azollae. Nitrogen fixing capacity and resistance to pests or ecological conditions of Azolla varies with species. The reasons for resistance are not known yet. Hence, cultivating algae-free Azolla and Anabaena azollae as well as recombining both of them would be useful for exploring the nature of root resistance, understanding the characteristics of the fern symbionts and the nature of the symbiosis. Probably it would also be a key point for successful breeding of Azolla. We have conducted experiments toward these goals since 1978. Algae-free Azolla was obtained by culturing the tip of the frond. Recombination of algae-free Azolla and Anabaena azollae was also attempted. Fifteen of 35 experiments proved to be successful. The frequency of recombination was about 43%. Recombination had occurred since the leaf cavities contained Anabaena azollae while algae-free Azolla did not. The nitrogen fixing capacity of recombining Azolla increased gradually. While essentially nothing is known about the process of recombination, our research efforts continued. The success of of recombination of algae-free <u>Azolla</u> and <u>Anabaena azollae</u> will open up new vistas for studying relationship between fern and <u>Anabaena azollae</u> and optimizing its practical application as a green manure in rice culture.

References

1 Chen Yang-chong, Zheng De-ying 1980 Use of <u>Azolla</u> spore technique. In: Information Office, Fujian Academy of Agr. Sc. (ed). Agric. Sc. Tech. Newsletter No. 5, 1-4.

2 Duan Bing-yuan, Zhang Zhuang-ta, Ke Yu-shi, Liu Xi-lian and Ling De-quan 1979 Studies on sexual propagation of <u>Azolla</u>. I. Effects of heredity and environmental conditions on sporogenesis. Guangdong Agr. Sc. No. 2, 23-27.

3 Duan Bing-yuan, Zhang Zhuang-ta, Ke Yu-shi, Liu Xi-lian and Ling De-quan 1979 Studies on sexual propagation of <u>Azolla</u>. II. Morphology of spore and the growing conditions required for sprouting and developing of <u>Azolla</u>. Guangdong Agri. Sc. No. 3, 54-57.

4 IRRI 1980 Annual report for 1980. The International Rice Research Institute, Laguna, Philippines. 263.

5 Lin Chang 1979 Effects of environmental conditions on growth of <u>Azolla</u>. Turang Feiliao No. 3, 36-39.

6 Liu Chung-chu, Lin Chang, Ren Zu-jian, and Lin Jia-keng 1979 Preliminary study on some physiological aspects of <u>Azolla</u>. Scientia Agricultura Sinica No. 2, 63-70.

7 Liu Chung-chu, Zheng De-ying, Chen Bing-huan, Chen Jia-ju, You Chong-Biao and Li Jiang-wei 1981 The potential of <u>Azolla</u> as a nitrogen source for paddy soil. <u>In</u>: A. H. Gibson and W. E. Newton (eds). Current Perspectives in nitrogen fixation. Australian Acad. of Sci. 501 p.

8 Liu Zhong-zhu, Chen Bing-huan and Song Wei 1981 Preliminary studies on nitrogen excretion by <u>Azolla</u>. <u>In</u>: Institute of Soil Science, Academia Sinica (ed). Proceedings of symposium on paddy soil. Science Press, Beijing, Springer-Verlag, Berlin, Heidelberg New York. 363-368.

9 Liu Chung-chu, Wei Wenshung, Zhin Guochang, Weng Beiqi and Zhang Yinqing 1982 Study on the potassium enriching physiology of <u>Azolla</u>. Scientia Agricultura Sinica. No. 4, 82:87.

10 Zheng Wei-wen, Lu pei-ji, Wei Wenshung 1982 Periodic changes of N_2-fixing activity of different species of <u>Azolla</u>. Fujian Agric. and Tech. No. 6, 22-24.

11 Yu Lin-da 1979 Preliminary observation on the sexual propagation of <u>Azolla</u>. Zhejiang Nongye Kexue. No. 4, 19-23.

12 You Shongbiao, Li Jingwei, Song Wei, Liu Chungchu, Wei Wenshung and Chen Binghuan 1981 Effects of N sources on some physiological characteristics of <u>Azolla</u>. Acta Phytophysiologia No. 2, 97-104.

Table 1: Quantity of N$_2$ fixed by various Azolla strains in different seasons and their relative values.

Kind of Azolla	Season	N$_2$ Fixed[a]			Night Relative to day	Night Relative to total
		Day	Night	Total		
		-- (mg of N/g Azolla) ---			-------- (%) --------	
A. caroliniana	Summer	0.103	0.027	0.131	27.0	21.2
	Autumn	0.215	0.155	0.370	71.9	41.8
	Winter	0.163	0.087	0.250	53.5	34.8
	Average				56.0	35.9
A. filiculoides	Summer	0.035	0.002	0.037	4.6	4.0
	Autumn	0.340	0.189	0.529	52.8	34.5
	Winter	0.159	0.045	0.204	28.6	22.0
	Average				42.4	29.8
A. imbricata (Xipin green)	Summer	0.008	0.000	0.008	14.1	12.4
	Autumn	0.400	0.049	0.449	42.7	29.9
	Winter	0.169	0.127	0.296	74.9	42.8
	Average				30.5	23.4

[a]Data are calculated from acetylene reduction method on the basis of Azolla fresh weight.

Table 2: Correlation between light, temperature and N_2 fixation.

Season	r or t [a]	A. caroliniana		A. filiculoides		A. imbricata [b]	
		LNF[c]	TNF[c]	LNF	TNF	LNF	TNF
Summer	r	0.449	0.476	0.650	0.360	0.850	0.720
	t	1.230	1.710	0.097	1.220	3.950**	3.280**
Autumn	r	0.460	0.840	0.830	0.887	0.860	0.900
	t	1.150	4.915**	3.320*	6.080**	3.760**	6.530**
Winter	r	0.060	0.922	0.250	0.938	0.212	0.930
	t	0.140	7.528**	0.447	8.560**	0.375	7.450**

a r = correlation coefficient
   t = test for significance of difference

b A. imbricata, Xipin green (Fujian)

c LNF = Light and N fixation
   TNF = Temperature and N fixation

* Significant

**Very significant

Table 3: Tolerance to shade of different Azolla strains.

Condition	Light Intensity and Length	Air Temperature (°C)	Water Temperature (°C)	Azolla Strain	Azolla Inoculation rate (g/pot)	Azolla Harvested (g/pot)
Under common water hyacinth	Mean Lux 21.3 k Direct light 3.5 hours	29.5	31.5	A[a] B[b] C[c]	3 3 3	3.50 1.90 4.25
Under leguminous plants	Mean Lux 6.3 k Direct light 1.5 hours	29.5	30.5	A B C	3 3 3	3.10 2.15 1.10
Under full grown rice plants	Mean Lux 3.5 k No direct light	28.8	30.0	A B C	3 3 3	0.38 0.00 0.00
No cover shade	Mean Lux 63 k Direct light 10 hours	31.3	33.5	A B C	3 3 3	3.10 3.45 5.75

a = A. caroliniana
b = A. imbricata, Xipin green
c = A. imbricata, Yulin

Table 4: Comparison of Azolla strains for resistance to thread worms (Polypedilum juinoense H)[a]

Azolla Strain	Azolla Weight (kg)		Weight Increase (%)	Thread Worm Count (Number/m²)		Increase in Number (%)
	Inoculated	Harvested		Beginning	Ending	
A. caroliniana	1.5	3.2	114	1,586	2,792	76
A. bengal serigor	1.5	2.7	81	965	2,774	187
A. imbricata, Yulin	1.5	2.2	47	1,719	6,351	270
A. imbricata, Xinpin green	1.5	2.6	76	772	6,249	709

a Experiment was conducted in a 2.5 m² pool.

Table 5: Growth rate of <u>A</u>. <u>caroliniana</u> compared with that of <u>A</u>. <u>imbricata</u> and <u>A</u>. <u>filiculoides</u>.

| Compared with | Increase in Rate of <u>A</u>. <u>caroliniana</u> | | | |
	Spring	Summer	Autumn	Winter
	------------------- (%) -----------------			
<u>A</u>. <u>filiculoides</u>	+0.97	_***	-9.1	+0.8
<u>A</u>. <u>imbricata</u> Xipin green	+16.2	+16.8*	+30.2	+106.7**

  * Significant
 ** Very significant
*** Very, very significant

5. Sporocarp germination, cytology and mineral nutrition of Azolla

P.K. Singh, R.N. Patra and S.K. Nayak
Laboratory of Blue-green Algae
Central Rice Research Institute
Cuttack-753006, Orissa, India

Key words Antimetabolites Azolla Cytology Growth regulator
Herbicides Mineral nutrition Pesticides Sporocarp germination

Summary

Azolla plants containing matured megasporocarps were collected during
the months of January to March, heaped for decomposition of vegetative
parts, sun dried and stored for germination study. After the dormancy
period of 3 to 5 months megasporocarps were inoculated and they
started germinating in pots after 7 days. Germination rate was
higher in the months of August to October. In field condition both
emergence and survival rates were fluctuating due to environmental
stresses. The somatic chromosome numbers (2n) were 40, 44, 48,
and 66 in A. filiculoides, A. pinnata (Indian isolate), A. mexicana
and A. pinnata (green isolate of Vietnam) respectively and later
one was found to be a hexaploid.
    The growth and nitrogen fixation were studied with reference to
macro and micro nutrients in Hoagland's medium under net house condition. The concentrations of elements supporting optimum growth
were as follows: P (2.5 to 15 pppm), Ca (10.0 to 20.0 ppm), Mg (30.0
ppm), K (15 to 45.0 ppm), Fe (2.0 to 5.0 ppm), Co (0.001 to 0.01
ppm) and Mo (0.01 to 0.1 ppm) where variation among the species
was observed. In the nutrient medium the effect of herbicides
(2,4-DNa, butachlor), pesticides (carbofuran, lindane), growth regulators (GA, IAA) and antimetabolites (chloramphenicol and DCMU) were
studied on the growth and N_2-fixation of Azolla where variations
in species were noticed.

1. Introduction

Azolla is cultivated through vegetative propogation. Due to reduced
viability in extreme conditions and its quick decomposible nature[9,16]
storage and transportation of the sporophyte become a problem.
The megasporocarp with its thick wall might tolerate these adversities
and therefore could be useful for long term germplasm preservation[5,17]
and convenient for transportation. However, detailed information
on methods of collection, preservation, its period of dormancy and
viability, the germination process and development of young sporophyte
is lacking. The chromosome number offers a useful taxonomic criteria
for distinguishing Azolla sps.[4,7,8]. Studies on effect of pesticides,
herbicides, growth regulators and antimetabolites on growth and
N_2-fixation of Azolla are also quite limited. The work briefly
reported here was conducted on sporocarp germination, cytology and
mineral nutrition during past 3 years.

2. Materials and Methods

2.1 Sporocarp collection and germination. The Azolla plants containing sporocarps were harvested and heaped for composting. Some
of the harvested plants were directly sun dried.   Harvesting of
sporocarps through the soil (mud) was also done where detached
megasporocarps were found to be numerous. The fine compost was
sieved whereas dried and sun baked soil was mixed thoroughly in
water in order to separate megasporocarps which floated on the water
surface. Healthy sporocarps of superior size were selected, surface
sterilized with 0.02% aquous solution of $HgCl_2$ for 3 min. and
inoculated into pots, beakers and petri dishes for germination studies.

2.2 Cytological studies. Shoot tips with young leaf primordia were
fixed in carnoy's fluid for 6 h., hydrolysed in 1:9 mixture of 1N HCl
and 2% aceto orcein at 60°C for 30 min. and finally squashed in
45% acetic acid.

2.3 Mineral nutrition, effect of biocides and growth regulators.
One isolate of A. filiculoides and A. mexicana and two isolates
of A. pinnata (Indian and green isolate of Vietnam) were used in
each experiment. Populations from a single vegetative plant of
each of the isolates was raised and maintained under exponetial
growth conditions. Cultures were maintained in 600 ml Corning glass
beakers containing 300 ml of 40% Hoagland's N-free culture medium[14]
with trace elements . These were incubated in wooden cages which
were fitted with wire gauge net (except the top and front which
was fitted with glass. In the net house the mean day/night temperature of 29°/21°C and sunlight illumination ranged from 12 to 26
klux. The cultures were transferred at 10 day intervals.
   For mineral nutrition studies Hoagland's solution with Allen and
Arnon's micro constituents was used except $MgCl_2$.$6H_2O$ replaced
$MgSO_4$.$7H_2O$ where the effect of Mg was studied. For biocides, growth
regulators and antimetabolites studies different doses of herbicide:
2,4-D, Butachlor; Pesticides:carbofuron, Lindane; growth regulators:
GA, IAA; antimetabolites:DCMU, chloramphenicol were used in N-free
nutrient medium to observe their effect on growth and N_2-fixation.

3. Results

3.1 Sporocarp germination. The sun dried megasporocarps did not
germinate even after 2 months of incubation. However, the sporocarps
collected from soil (Figure 1) and Azolla compost germinated (Figure
2) after 5 to 6 months of harvest when incubated in small earthen
or ceramic pots with flooded soil kept under net house condition
at 60 to 80 klux light intensity. These sporocarps did not germinate
when incubated in sterilized petri dishes containing moist blotting
paper with medium or in beakers with nutrient medium. Similar attempt
to germinate in complete dark was a failure, whereas in shade at
low light intensity of 0.5 to 2.0 klux, there was 1 to 3% emergence
but seedlings did not survive further. The multiplication plots
which were alternately dried and wetted after sporulation for 5
to 6 months showed young seedlings within 13 days of irrigation.
The soil of multiplication plots contained lesser number of

megasporocarps in comparison to Azolla compost prepared after sporulation (Table 1). The total yield of megasporocarps measured in
0.33 m tray in a sporulating season was higher in case of A. mexicana
followed by A. pinnata (Bangkok isolate) (Table 1). The dry weight
of 100 sporocarps of A. mexicana was less than that of the strains
of A. pinnata studied. The sporocarps which were given a heat treatment at 45°C in an incubator for 7, 15 and 30 days germinated after
5 to 6 months of storage. In all isolates of Azolla studied germination took place within 10 ± 3 days of incubation both under net
house and field conditions. The germination rate was less in field
condition inoculated in 1.0 m plot where previously no Azolla had
been grown (Table 2). The survival rate was also more in net house
condition (Table 3).

Megasporocarps-floats looked like silver shining patches and signs
of germination was evident when these patches were peeled off from
the supraspore region. The supraspore region on the other hand,
turned greenish with anostomising threads prior to emergence of
the cotyledon. A deeply notched and funnel shaped cotyledon (Figure
3) emerged pushing the dark tan cloured indusium cap to one side.
Below the cotyledon the first root came out just above the cap within
2 days of germination while the second root emerged in between the
first and second leaf within a week at which time the first root
had already attended its full growth. As regards to symmetry, the
long axis of megasporocarp was parallel to the water surface. After
the appearance of cotyledon the axis turned oblique to the water
surface. The shoot apex was slightly bent, grew dorsiventrally
on the water surface and growth was slow up to the 15-18 leaf stage.
The third root appeared to be the indicator of further growth and,
hence, survival. At cotyledonary stage some seedlings also perished.
In A. mexicana empty spore cases generally get detached early but
in A. pinnata (India, Nepal, Bangladesh and Thailand isolates) it
remained attached even up to full sporophytic stage provided there
was no mechanical injury. At the 15 leaf stage the cotyledon started
weathering gradually but remained persistent. The full sporophytic
stage was attained in 30 ± 5 days of incubation. Anabaena resting
cells were observed in megasporocarps and its isolated thick walled
cells were noticed in early stage of first leaf, whereas in mature
stage of first leaf cyanobacterial filaments were seen. The fully
unfurled cotyledons of A. pinnata were larger and had more stomata
on the inner faces of the funnel compared to those of A. mexicana.

3.2 Cytology. The nuclei in most cases were observed to be oval
and in some cells, elliptical. The nucleolus was not marked in
any cell. The chromosomes at premetaphase were well condensed,
short and rod shaped. The somatic chromosome numbers (2n) were
countable to 40 in A. filiculoides, 44 in A. pinnata (India) 48
in A. mexicana and 66 in A. pinnata (green isolate of Vietnam).
The centromere was not traced in any case due to the extreme small
size of chromosomes. At mid metaphase the chromosomes were aggregated
at the equatorial plate loosing their identity. At anaphase spindle
fibre was freely observed. Precocious movement of certain chromosomes
was marked in separation. The daughter cells which formed after
telophase did not start simultaneous division. Both Indian and
Vietnam isolates of A. pinnata showed a common basic number
i.e. x = 11, where the later isolate was observed to possess superior

morphological features (Table 4).

3.3 Mineral nutrition and the effects of herbicides, pesticides, growth regulators and antimetabolites. In general the growth of Azolla in the standard N-free medium was very rapid. An inoculum of 0.3 g fresh weight of Azolla in 300 ml medium produced 3-5 g of fresh matter in ten days. An increase in total N-content from 1.82 to 6.52 mg under similar conditions showed that Azolla fixes atmospheric nitrogen efficiently. The generation time of Azolla varied from 2.5 to 4.5 days. The effects of various macro-and micro-nutrients (Table 5), pesticides, herbicides, growth regulators and antimetabolites (Table 6) are given below:

3.4 Mineral nutrition. Phosphorus. P deficiency caused less growth, decreased frond size, browning of lobes with long unhealthy roots and after pre-longed starvation some fronds died. The micro symbiont was faint in colour and the heterocyst frequency was reduced. The P requirement was studied at 2.5, 10.0 and 15.0 ppm. In case of A. filiculoides the growth and nitrogen fixation increased with the increase of P levels till 15.0 ppm of P. The nitrogen fixed at 10.0 ppm and 15.0 ppm was 6.96 mg and 7.11 mg respectively with generation time of 2.77 days in both concentrations. In case of A. mexicana there was significant increase in growth and nitrogen fixation with the increase of P levels up to 15.0 ppm, producing a maximum fresh weight of 3.12 g, a dry weight of 136.67 mg, a genera-tion time of 2.96 days and nitrogen fixation of 6.46 mg at 15.0 ppm. In A. pinnata (India) the increase in growth and nitrogen fixation was found up to 10.0 ppm producing fresh weight of 3.62 g, dry weight of 160.83 mg, generation time of 2.78 days and nitrogen fixation of 7.58 mg then decrease in rate was observed at 15.0 ppm. In A. pinnata (Vietnam) growth and nitrogen fixation increased up to 5.0 ppm producing a fresh weight of 3.65 g, a dry weight of 172.5 mg, a generation time of 2.77 days and nitrogen fixation of 9.11 mg. A decrease was noticed with increased P levels.

3.5 Calcium. Deficiency of calcium in the medium caused a dark browning of the fronds in the central lobes with an increase in anthocyanin pigments after 6-7 days. After long starvation the plants became necrotic and frond size decreased. The calcium require-ment was studied at 0.0, 10.0, 20.0 and 30.0 ppm. The increase in growth and nitrogen fixation of A. filiculoides was noticed up to 10.0 ppm. In A. mexicana the growth and nitrogen fixation in-creased from 0.0 to 20.0 ppm like A. pinnata (India), but in A. pinnata (Vietnam) the growth and nitrogen fixation increased up to 20.0 ppm. A. filiculoides responded better at all levels of Ca than other species/varieties.

3.6 Magnesium. The effect of Mg was studied at concentrations of 0.0, 15.0, 30.0 and 50.0 ppm. A. filiculoides showed optimum growth and N_2-fixation up to 30.0 ppm of Mg where the fresh weight was 3.2 g, the dry weight was 151.7 mg, 7.87 mg of N was fixed and the generation time was 2.93 days. These parameters decreased at higher levels of Mg. In case of A. mexicana optimum growth and nitrogen fixation were observed at 30.0 ppm of Mg where the fresh weight was 2.53 g, the dry weight 116.7 mg, the generation time 3.25 days

and 6.58 mg of N was fixed. In A. pinnata (India and Vietnam isolates) optimum growth and nitrogen fixation also occurred at 30.0
ppm.

3.7 Potassium.   Potassium deficiency over concentrations of 0, 15,
45 and 90 ppm also had no visible symptom like Mg, although after
long starvation there was a decrease in growth and nitrogen fixation.
A. filiculoides was affected adversely at zero level of K and there
was increase in growth up to 15.0 ppm.   There was increase in both
growth and N_2-fixation up to 45.0, 15.0 and 45.0 ppm in A. mexicana,
A. pinnata (India) and A. pinnata (Vietnam) respectively.

3.8 Iron.   In iron deficiency, plants turned yellowish with decreased
frond size and browning of fronds as was observed in P deficiency.
Different levels of iron (0.0, 1.0, 2.0 and 5.0 ppm) as Fe-EDTA
were studied with four Azolla isolates.   A. filiculoides required
iron for growth and N_2-fixation which increased at levels up to
5.0 ppm producing a dry weight of 105.0 mg and N_2-fixation of 4.94
mg at this level.   In A. mexicana, A. pinnata (India) and A. pinnata
(Vietnam) growth and N_2-fixation increased up to 2.0 ppm producing
dry weight of 129.0 mg, 186.67 mg, 166.67 mg and N_2-fixation of
7.75 mg, 11.47 mg and 8.82 mg respectively at this level.   In comparison A. pinnata (India) responded better and A. filiculoides poorer
among species/varieties used in experiments.

3.9 Molybdenum.   Deficiency of the micronutrient Mo did not produce
any morphological change but decreased the growth and N_2-fixation.
Different levels of Mo (0.0, 0.001, 0.01 and 0.1 ppm) were studied
with four Azolla isolates.   In A. filiculoides and A. pinnata
(Vietnam) growth and N_2-fixation increased up to 0.01 ppm producing
dry weight of 130.0 mg, 135.0 mg and N_2-fixation of 6.29 mg and
9.07 mg respectively.   But in A. mexicana and A. pinnata (India)
growth and N_2-fixation increased up to 0.001 ppm producing dry weight
of 107.5 mg, 131.7 mg and N_2-fixation of 4.47 mg and 5.32 mg respectively.

3.10 Cobalt.   Cobalt deficiency restricted growth and N_2-fixation
as noticed for molybdenum.   Different levels of cobalt (0.0, 0.001,
0.01 and 0.1 ppm) were studied with four Azolla isolates.   In case
of A. filiculoides, A. mexicana and A. pinnata (Vietnam) growth
and N_2-fixation increased up to 0.01 ppm whereas in A. pinnata (India)
growth and N_2-fixation increased up to 0.001 ppm of Co.   Further
increase in concentration decreased both growth and N_2-fixation.

3.11 Pesticides effects

3.11.1 Carbofuran.   Different levels of carbofuran (0.0, 0.25, 0.50
and 1.0 ppm) were studied on Azolla sps.   The growth and N_2-fixation
increased with the application of lower doses of carbofuran to Azolla.
In A. filiculoides and A. pinnata (Vietnam) growth and N_2-fixation
increased up to 0.5 ppm producing dry weight of 120 mg, 107 mg and
N_2-fixation of 7.32 mg and 5.82 mg respectively.   But the growth
and N_2-fixation increased up to 0.25 ppm of carbofuran in A. mexicana
and A. pinnata (India).   Further increase in carbofuran concentration
up to 1.0 ppm decreased fresh weight, dry weight, generation time

and N_2-fixation in all the Azolla isolates.

3.12 Lindane. Different levels of lindane (0.0, 0.01, 0.1 and 0.25 ppm) were studied on Azolla sps. Addition of lower oncentrations of lindane also increased growth and N_2 -fixation in Azolla. In A. filiculoides, A. pinnata (India) and A. pinnata (Vietnam) growth and N_2 -fixation increased up to 0.01 ppm producing dry weight of 122.5 mg, 115.84 mg, 108.34 mg and N_2 -fixation of 7.47 mg, 7.13 mg and 6.02 mg respectively. But in A. mexicana growth and N_2-fixation increased up to 0.1 ppm of lindane producing 105.0 mg dry weight and 5.57 mg N_2 -fixation. Further increase of concentration up to 1.25 ppm decreased growth and N_2-fixation in Azolla isolates.

3.13 Herbicides effect

3.13.1 2,4-D. Different levels of 2,4-D (0.0, 0.05, 0.10 and 0.50 ppm) were studied with four Azolla isolates. The herbicide 2,4-D also increased growth and N_2-fixation. In A. filiculoides the growth and N_2 -fixation increased up to 0.05 ppm producing dry weight of 98.34 mg and N_2 -fixation of 4.49 mg then decreased with increase in concentration of 2,4-D up to 0.5 ppm. In A. mexicana, A. pinnata (India) and A. pinnata (Vietnam) growth and N_2 -fixation increased up to 0.1 ppm of 2,4-D producing dry weight of 127.5 mg, 155.0 mg, 150.0 mg and N_2 -fixation of 6.84 mg, 7.82 mg and 8.61 mg respectively. In comparison better results were obtained in A. pinnata India and Vietnam whereas poor response was noticed in A. filiculoides for growth and N_2-fixation.

3.14 Butachlor. Different levels (0.0, 0.001, 0.0025 and 0.01 ppm) of butachlor were studied with four Azolla isolates. The lower doses of butachlor stimulated growth and N_2-fixation of Azolla. In A. filiculoides, A. mexicana and A. pinnata (India) 0.001 ppm of butachlor increased growth and N_2 -fixation producing dry weight of 105.84 mg, 90.84 mg, 116.5 mg and N_2 -fixation of 5.36 mg, 4.36 mg and 5.94 mg respectively, then growth and N_2 -fixation decreased with increase of butachlor up to 0.01 ppm. But in the case of A. pinnata (Vietnam) which showed resistance to butachlor in comparison to others increased growth and N_2-fixation up to 0.0025 ppm, producing dry weight of 100.84 mg and N_2-fixation of 4.59 mg. A. mexicana was found susceptible to butachlor in comparison to others.

3.14 Effect of growth regulators

3.14.1 Gibberllic acid. Different levels (0.0, 0.001, 0.1 and 5.0 ppm) of GA were studied on Azolla sps. The effect of gibberllic acid (GA) on the growth and N_2 -fixation of Azolla was not very pronounced. In A. filiculoides, A. pinnata (India) and A. pinnata (Vietnam) 0.001 ppm increased growth and N_2-fixation producing 143.34 mg, 140.0 mg, 141.67 mg of dry weight and 7.25 mg, 6.8 mg and 7.67 mg of N_2-fixation respectively, then growth and N_2-fixation decreased with increase of GA up to 5.0 ppm. But in A. mexicana growth and N_2-fixation increased up to 0.1 ppm producing dry weight of 124.17 mg and nitrogen fixation of 5.7 mg, then growth and N_2-fixation decreased at its higher levels.

3.15 Indole acetic acid. Effect of growth regulator IAA of different concentrations (0.0, 0.01, 0.1 and 1.0 ppm) was studied with four _Azolla_ isolates. Growth and N_2-fixation in _A. filiculoides_, _A. mexicana_ and _A. pinnata_ (India) increased at 0.01 ppm producing dry weight of 123.34 mg, 99.17 mg and N_2-fixation of 6.14 mg, 4.48 mg and 5.91 mg respectively, then decreased with the increase of IAA up to 1.0 ppm. But in case of _A. pinnata_ (Vietnam) growth and N_2-fixation increased up to 0.1 ppm of IAA producing dry weight of 127.5 mg and N_2-fixation of 5.35 mg, then decreased at 1.0 ppm.

3.16 Effect of antimetabolites

3.16.1 DCMU. Different levels of DCMU (0.0, 0.001, 0.005 and 0.01 ppm) were studied. The DCMU was also found to act as a growth stimulatory substance at lower dose. _A. filiculoides_, _A. mexicana_ and _A. pinnata_ (India) increased growth and N_2-fixation at 0.001 ppm of DCMU producing dry weight of 178.35 mg, 168.34 mg, 197.0 mg and N_2-fixation of 8.97 mg, 8.56 mg and 9.47 mg respectively. But in _A. pinnata_ (Vietnam) growth and N_2-fixation decreased at all the levels of DCMU used in experiment. _A. mexicana_ was found to be tolerant and _A. pinnata_ (Vietnam) was found to be susceptible to this chemical.

3.17 Chloramphenicol. Different levels (0.0, 0.01, 0.1 and 1.0 ppm) of this antibiotic were studied with four _Azolla_ isolates. In _A. filiculoides_, _A. mexicana_ and _A. pinnata_ (India) growth and N_2-fixation increased at 0.01 ppm producing dry weight of 112.15 mg, 160.84 mg, 196.67 mg and N_2-fixation of 5.98 mg, 8.51 mg and 10.86 mg respectively. Further increase in concentration of chloramphenicol decreased growth and N_2-fixation. But in _A. pinnata_ (Vietnam) growth and N_2-fixation increased up to 0.1 ppm producing dry weight of 204.0 mg and N_2-fixation of 11.02 mg, then growth and N_2-fixation decreased with the increase of chloramphenicol up to 1.0 ppm. _A. pinnata_ (Vietnam) was found to be comparatively resistance to this chemical.

4. Discussion

The smaller mass of megasporocarps and their tolerance to adverse conditions prompted us to study its germination so that it may be transported easily without loss of viability. A dormancy period of 4-5 months was observed in _A. pinnata_ and 2-3 months in _A. mexicana_ suggesting earlier fertilization and embryo development in the latter. However, the details of the fertilization process is still unknown in _Azolla_. Sun baking might be an important factor for breaking the dormancy but dry storage does not appear to be a factor as megasporocarps sun baked for 6 months without dry storage also germinated. On the other hand dry storage for longer periods i.e. for one year or more was injurious to the embryo since megasporocarps stored for one year in the cup-board did not germinate. Therefore, long term preservation of _Azolla_ germplasm may not be possible through dry storage. Turrill[20] reported that seeds remain viable for years if buried in soil. Hence it is likely that sporocarps of _Azolla_ remain viable in mud of ponds and ditches and germinate when favorable period prevails. Water storage might also help keep aquatic spores viable. At a water temperature of 30 ± 6°C the healthy non-dormant megasporocarps readily germinate within 7 days of incubation in flooded soil both in net house and field condition. Fu et al.[5] observed sporocarps germinated after 20 days of incubation in China. The megasporocarps of Azolla appears to be sun loving spores since in bright sun-light (90 to 110 klux light intensity) 70 to 100% emergence of cotyledons occurred. Under field condition germination of inoculated sporocarps was less due to disturbance by wind, rainfall and fluctuation in water level. Occurrence of algae also affected adversely the growth of young seedlings. Thus, Azolla cultivators are advised to germinate megasporocarps in small containers and transfer to another container in order to increase the density before transporting to the field.

The variation in chromosome number clarified the status of the three species. Basing on present findings the Indian isolate and Vietnam green isolate of Azolla cannot be grouped into two different species although these show a 44 and 66 chromosome number, respectively. Rather these can be interpreted as two different cytotypes of common species i.e. Azolla pinnata having basic number x = 11. Bir[2] classified A. pinnata (India) as tetraploid. When the Vietnam green isolate of Azolla was compared with the diploid cytotype (i.e. the Indian isolate both morphologically and ecologically). The former showed superior characters and greater ecological tolerance which might be due to higher ploidy level i.e. hexaploid. Whether the origin of polyploidization in A. pinnata is recent or primitive cannot be concluded at this stage and therefore further study on meiosis is suggested.

Long term deficiency of P, Ca and Fe affected Azolla growth more adversely in comparison to Mg, K, Mo and Co in this study. The fresh matter produced by the Vietnam green isolate of A. pinnata was more than the other species/varieties with P deficiency as well as optimum doses of P. This might be due to its high absorption ability. Similar trends were observed in dry matter and nitrogen fixation. For the healthy growth and nitrogen fixation our observations show that Azolla requires widely differing 'P' concentrations in the culture medium.

Ca deficiency caused dark browning of fronds in 5 days, necrosis after long starvation, decrease in frond size, increase in anthocyanin pigments and some times loss of Anabaena azollae[9,18,21], A. filiculoides and A. pinnata (Vietnam) showed better response to calcium than A. mexicana and A. pinnata (Indian isolate) but A. filiculoides fixed more nitrogen/beaker than others at the minimum level of Ca. High doses of calcium (> 20 ppm) decreased the growth and nitrogen fixation in Azolla sps.

Effect of Mg on growth and nitrogen fixation was less pronounced than that of P, Ca, K as reported by other workers[18,23]. The reduction in fresh weight to 82% and total nitrogen content to 77% was reported[22]. A. filiculoides showed better response to growth and nitrogen fixation at the optimum dose of Mg than other species of Azolla and poor response was observed in A. mexicana and A. pinnata (Vietnam).

Although K deficiency had no visual symptoms like that of P and Ca, A. pinnata (India) and A. mexicana showed better response than A. filiculoides and A. pinnata (Vietnam). At the optimum dose of 15.0 ppm K generation time was minimum in A. pinnata (India).

Iron is required by Azolla and its availability depends on the pH of the media. It is used in the media with chelating agents like EDTA and citrate for its better availability. At low pH iron occurs in ferrous forms and might not be utilized by Azolla but Azolla utilizes the ferric form [11]. A. pinnata (India and Vietnam isolates) showed a better response in both Fe deficiency and Fe supplemented media for growth and N_2-fixation than A. filiculoides and A. mexicana.

The micronutrient requirement of Azolla is similar to the micro-nutrients required by blue-green algae[1] and therefore the micro-nutrients used in Allen and Arnon[1] algal medium has been[14,16] also used for Azolla growth in the Hoagland's medium[11]. Olsen[11] observed that the addition of 0.05 ppm of Mo in solution caused a 64% increase in N_2-fixation. In the present observation Mo (0.001-0.1 ppm) favoured A. pinnata (Vietnam) and A. filiculoides growth and N_2-fixation better than A. mexicana and A. pinnata (India). Bortels[3] found 83% increase of N_2-fixation by the addition of 0.01 ppm of Mo to Azolla culture. Johnson et al[6] showed that 0.01 µg Co/L supplied as $CoCl_2$ to A. filiculoides cultures growing in a N-free medium resulted in an increase of chlorophyll and N_2-fixation as compared to the treatment without Co. In the present finding A. pinnata (India and Vietnam isolates) showed better response to Co than A. filiculoides and A. mexicana and the generation time was minimum in A. pinnata (India) at 0.001 ppm of Co. From the growth rate and N_2-fixation analysis it is clear that the macro and micronutrients are essential for Azolla sps.

Sometimes the growth of Azolla is reported to be affected adversely by the application of herbicides. The herbicides at lower doses were found to encourage the growth and N_2-fixation. A. pinnata (India and Vietnam isolates) showed better response than A. filiculoides and A. mexicana. A. filiculoides were susceptible to 2,4-D whereas A. pinnata (Vietnam) was found to be comparatively resistant. Herbicides acted like hormone at lower doses as a result of which rhizome got elongated with wide spacing in between leaves although at higher doses of 2,4-D, fronds became fragile, fragmented into smaller pieces and roots became detached. 2,4-D is also known[15] to encourage the growth of blue-green algae. High doses of DCMU (0.005 to 0.01 ppm) decreased the growth and nitrogen fixation. Chloramphenicol at its low doses of 0.01 to 0.1 ppm increased the growth and N_2-fixation of A. pinnata (Vietnam and Indian isolates) whereas decreasing trend was noticed in A. mexicana and A. filiculoides at 0.1 ppm, although higher concentration of 1.0 ppm was inhibitory to all of them. Peters reported that chloramphenicol effects N_2-fixation. In the present study the effect was studied after incubation for 10 days. It would be desirable to study the effects of antimetabolites in detail immediately after their addition so that their primary effect could be ascertainal.

Acknowledgments We gratefully acknowledge Dr. H. K. Pande, Director, Central Rice Research Institute, Cuttack for providing the necessary facilities and encouragement.

References

1 Allen M B and Arnon D I 1955 Studies on nitrogen fixing

blue-green algae. I. Growth and nitrogen fixation by _Anabaena cylindrica_ Lemm. Pl. Physiol. 30: 366-372.

2 Bir S S 1973 Cytology of Indian Pteridophytes. Reprinted from Glimpses in Plant Research Vol I pp 28-119.

3 Bortels H 1940 Uber die bedentung des molybdans fur die Sticksoffnindenden Nostocean Arch. Mikrobiol. 11: 155-186.

4 Ducan R E 1940 The cytology of sporangium development in _Azolla filiculoides_. Bull. Torry Bot. Club. 67: 391-412.

5 Fu Cheng Jing, Xi Sheng-Xis and Wang Su Zhen 1978 The sporocarps and cultured young sporophytes. Acta Botanica Sinica 20: 54-58.

6 Johnsan G V, Mayeux P A and Evans H J 1966 A cobalt requirement for symbiotic growth of _A. filiculoides_ in the absence of combined nitrogen. Pl. Physiol. 4: 852-855.

7 Loyal D S 1958 Cytology of two species of salviniacease. Curr. Sci. 27: 357-358.

8 Loyal D S 1972 Chromosome size and structure in some heterosporous ferns _In_ Advancing frontier in cytogenetics and improvement of plants. Ed. P.N. Kachroo. Hindustan Publication, New Delhi. pp. 293-298.

9 Moore A W 1969 _Azolla_, biology and agronomical significance. Bot. Rev. 35: 17-34.

10 Nickel L G 1961 Physiological studies with _Azolla_ under aseptic conditions. II Nutritional Studies and effect on growth. Phyton (Buenos Aires) 17: 49-54.

11 Olsen C 1970 On biological nitrogen fixation in nature in blue-green algae. Compt. Rend. Trav. Lab. Carlsberg. 37: 269-283.

12 Peters G A 1975 The _Azolla-Anabaena azollae_ relationship. III studies on metabolic capabilities and a further characterization on the symbiont. Arch. Mikrobiol. 103: 113-122.

13 Peters G A 1976 Studies on the _Azolla-Anabaena azollae_ symbiosis. _In_ Proceedings of the 1st international symposium on nitrogen fixation. Eds Newton W E and Nyman C J, State University Press, Pullman. pp. 592-610.

14 Peters G A and Mayne B C 1974 The _Azolla-Anabaena azollae_ relationship. I. Initial characterization of the association. Pl. Physiol. 53: 3813-3819.

15 Singh P K 1974 Algicidal effect of 2,4-D on blue-green algae _Cylinderospermum_ sps. Arch. Microbiol. 77: 69-72.

16 Singh P K 1979 Symbiotic algal N_2-fixation and crop productivity. _In_ Annual Review of Plant Sciences. Ed. C P Mallik. Vol I. Kalyani Publishers, New Delhi. pp. 1-22.

17 Singh P K, Satapathy K B, Misra S P, Nayak S K and Patra N 1982 Application of _Azolla_ in rice cultivation. _In_ Biological Nitrogen Fixation. Bhabha Atomic Research Centre, Trombay, Bombay. pp. 423-450.

18 Subudhi B P R and Singh P K 1979 Effect of macronutrients and pH on the growth, nitrogen fixation and soluble sugar content of water fern _Azolla pinnata_. Biol. Plant. 21: 66-70.

19 Subudhi B P R and Watanabe I 1981 Differential phosphorus requirement of _Azolla_ sps. and strains in phosphorus limited continuous culture. Soil Sci. Plant. Nutri. 27: 237-247.

20 Turrill W B 1957 Germination of seeds: 5. The vitality and longivity of seeds. Ganrs. Chron. 142: 37-48.

21 Watanabe I 1978 _Azolla_ and its use in lowland rice culture. Soil and Microbe (Japan) 20: 1-10.

22 Watanabe I, Espinas C R, Berja N S and Alimagno B V 1977 Utilization of the Azolla-Anabaena complex as a nitrogen fertilizer
for rice. International Rice Research Institute. Paper Series.
11, pp. 1-15.
23 Yatazawa M, Tanatsu N, Hosoda N and Nuname K 1980 Nitrogen
fixation in Azolla-Anabaena symbiosis as affected by mineral
nutrient status. Soil Sci. Plant Nutr. 26: 415-426.

Table 1: Average number, dry weight and percent of fertile megasporocarps in soil of different isolates of Azolla.

Azolla	Number of sporocarps/ 100 g of soil	Number of sporocarps/ 100 g of Azolla compost	% fertile sporocarps in soil	% fertile sporocarps in Azolla compost	Dry weight of 100 megasporocarps (in mg)	Fresh weight of both micro and megasporocarps/ g of sporophyte (in g)	Total yield of megasporocarps/ 0.33 m² tray
Azolla mexicana (USA)	1500	3000	54.6	21.6	0.14	0.32	14.2
Azolla pinnata (India)	1200	2500	43.1	15.3	0.28	0.15	8.6
Azolla pinnata (Thailand)	800	1600	40.0	12.6	0.17	0.26	12.0
Azolla pinnata (Nepal II)	1100	2100	38.0	8.6	0.27	0.23	8.4
Azolla pinnata (Bangladesh)	1100	1800	42.6	16.0	0.25	0.13	6.8
Azolla pinnata (Africa)	200	400	46.3	18.0	0.27	0.10	-

- Not determined

Table 2: Emergence (%) of sporocarps collected through soil, incubation in the field and in containers placed under net house condition.

Azolla	Condition	1981* July	August	September	October	November	December	1982** July	August	September	October
Azolla mexicana	Field	–	–	8.6	5.3	8.0	–	7.6	4.0	5.0	4.6
	Net house	–	29.0	48.0	42.6	43.6	12.6	18.3	55.0	48.3	53.0
A. pinnata (India)	Field	–	9.6	8.0	7.0	8.3	–	10.3	7.6	2.6	8.3
	Net house	–	40.0	96.3	73.6	32.6	18.3	25.6	64.0	76.3	71.6
A. pinnata (Bangladesh)	Field	–	–	–	–	–	–	8.3	8.0	7.6	5.6
	Net house	–	–	–	–	–	–	48.6	51.6	57.6	52.3
A. pinnata (Thailand)	Field	–	5.6	4.3	5.0	2.6	–	4.6	5.3	5.0	6.3
	Net house	–	10.0	57.0	54.6	24.0	15.0	34.3	28.3	31.0	42.6
A. pinnata (Nepal II)	Field	–	–	–	–	–	–	7.3	6.6	7.3	6.6
	Net house	–	–	–	–	–	–	68.6	62.3	59.0	47.6
A. pinnata (India) (One year storage)	Field	–	0	0	0	0	0	0	0	0	0
	Net house	–	0	0	0	0	0	0	0	0	0

0 No germination

* Sporocarps harvested in 1980–81 season

** Sporocarps harvested in 1981–82 season

– Not determined

67

Table 3: Percent survival of young seedlings after 30 days of incubation of megasporocarps of _Azolla_.

Azolla	Condition	1981					1982			
		August	September	October	November	December	July	August	September	October
Azolla mexicana	Field	–	27.7	24.5	20.0	–	30.2	25.0	40.0	21.7
	Net house	70.0	80.4	82.2	87.2	84.1	87.4	87.3	90.3	85.0
A. pinnata (India)	Field	34.4	78.8	71.4	48.1	–	19.4	34.2	0.0	24.1
	Net house	87.5	77.9	91.0	82.8	18.3	84.4	82.8	91.7	24.0
A. pinnata (Bangladesh)	Field	–	–	–	–	–	21.7	25.0	17.0	17.9
	Net house	–	–	–	–	–	87.7	83.9	69.4	72.7
A. pinnata (Thailand)	Field	17.9	83.5	20.0	0.0	–	21.7	35.8	20.0	31.7
	Net house	60.0	80.0	76.9	83.3	86.6	87.5	72.8	90.3	79.8
A. pinnata (Nepal II)	Field	–	–	–	–	–	31.5	15.2	13.7	19.7
	Net house	–	–	–	–	–	85.0	82.9	82.4	84.0

– Not determined

Table 4. Comparative feature of Indian and Vietnam green isolate of _Azolla_ _pinnata_ (grown in field condition).

Features	Indian isolate	Vietnam (Green) isolate
Frond size (mm)	249.7	382.4
Shoot length (mm)	17.2	18.6
Root length (mm)	50.4	80.3
Shoot-trichome length (μm)	132	154
Root hairs during mat formation (mm)	2.2	3.0
Leaf colour	Purple	Always green
Size of dorsal lobe (mm)	1.39/0.63	1.53/0.69
Size of ventral lobe (mm)	1.44/1.20	1.52/1.55
Stomatal size (length/breadth in μm)	166.8/45.7	193.7/59.1
Microsporocarp (length/breadth in mm)	2.13/1.6	2.33/1.6
Megasporocarp (length/breadth in mm)	0.76/0.48	0.79/0.48
Microsporogia/sporocarp	70	135
Glochidia size (length/breadth in μm)	123.7/35.0	150.1/37.7
Tolerance to high temperature	Susceptible	Tolerant
Somatic chromosome number	44	66

Table 5. Optimum levels of macro and micronutrients supporting best N_2-fixation in <u>Azolla</u> sps.

Nutrients	Concentration (ppm)			
	<u>A</u>. filiculoides	<u>A</u>. mexicana	<u>A</u>. pinnata (India)	<u>A</u>. pinnata (Vietnam)
P (KH_2PO_4)	15.0 (7.10)	15.0 (6.46)	10.0 (7.58)	5.0 (9.11)
Ca ($CaCl_2.2H_2O$)	10.0 (8.06)	20.0 (6.36)	10.0 (6.28)	20.0 (7.38)
Mg ($MgCl_2.6H_2O$)	30.0 (7.87)	30.0 (6.58)	30.0 (6.48)	30.0 (6.29)
K (KCl)	15.0 (6.60)	45.0 (7.09)	15.0 (7.83)	15.0 (6.81)
Fe (Fe-EDTA)	5.0 (4.94)	2.0 (7.75)	2.0 (11.47)	2.0 (8.82)
Mo ($Na_2Mo_3.2H_2O$)	0.01 (6.29)	0.001 (4.47)	0.001 (5.32)	0.01 (6.07)
Co ($CoCl_2.6H_2O$)	0.01 (1.53)	0.01 (9.67)	0.001 (14.72)	0.01 (11.95)

Figures in parentheses indicate total mg N_2 fixed/300 ml/beaker at 10th day of incubation.

Table 6:   Effect of pesticides, herbicides, antimetabolites (non-inhibitory concentrations) and growth regulators (growth stimulatory concentrations) on N_2-fixation of A̲z̲o̲l̲l̲a̲ sps.

| | Concentrations (ppm) | | | |
	A̲. filiculoides	A̲. mexicana	A̲. pinnata (India)	A̲. pinnata (Vietnam)
Pesticides				
Carbofuran	0.50 (7.32)	0.50 (3.83)	0.50 (4.76)	0.50 (5.82)
Lindane	0.10 (6.22)	0.10 (5.57)	0.10 (7.13)	0.10 (5.13)
Herbicides				
2,4-D	0.10 (4.24)	0.50 (5.74)	0.10 (7.82)	0.10 (8.61)
Butachlor	0.01 (4.41)	0.0025 (3.56)	0.0025 (4.62)	0.01 (3.74)
Antimetabolites				
DCMU	0.001 (8.97)	0.005 (6.49)	0.01 (6.56)	0.0 (9.46)
Chloramphenicol	0.01 (5.98)	0.01 (8.51)	0.10 (10.37)	0.10 (11.02)
Growth Regulators				
GA	0.001 (7.25)	0.1 (5.70)	0.001 (6.80)	0.001 (7.67)
IAA	0.01 (6.14)	0.01 (4.48)	0.01 (5.91)	0.1 (5.35)

Figures in parentheses indicate mg N_2 fixed/300 ml/beaker

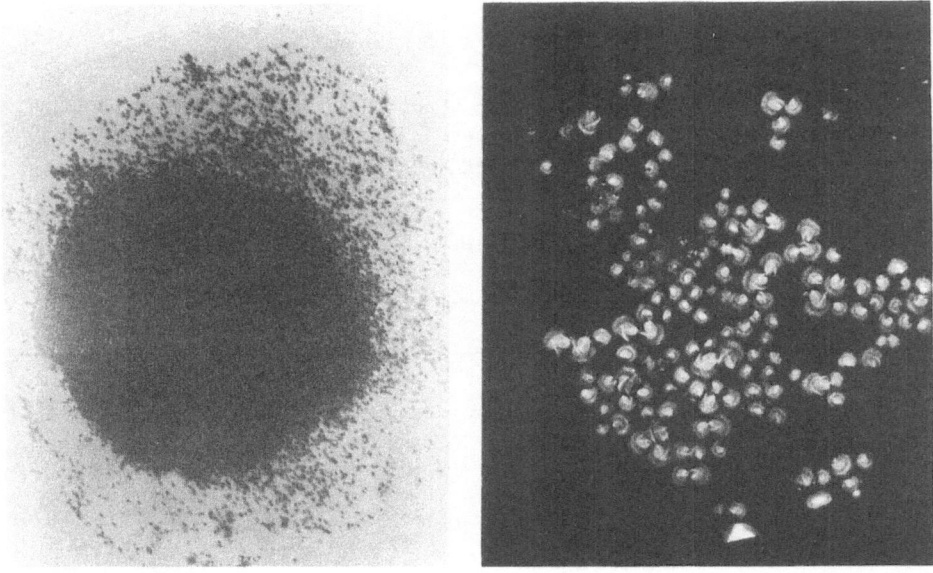

Figure 1. Purified megasporocarps of A. pinnata

Figure 2. Germinating megasporocarps

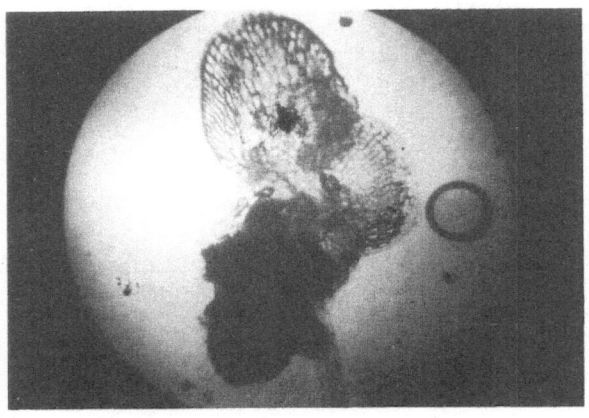

Figure 3. Enlarged view of a germinated megasporocarp

6. Azolla pinnata var. Africana: Background, ecophysiology, plots assays

P. A. Reynaud
ORSTOM
B.P. 1386
Dakar, Sénégal

Key words   Azolla pinnata var. Africana   Ecophysiology   Humidity Light Nitrogen Sénégal Temperature

Azolla pinnata variety africana was first described by Desvaux in 1827 as follows: a floating aquatic fern with deltoids fronds, a main step up to 2 cm long horizontal, minutely papillate, leaves 2 lobed, green to reddish, ovate, no scarious, repeatedly alternately branched, macrospores surrounded by numerous "flotation bodies", granular and microspores with numerous massulaes.

1. Background

According to a number of recent bibliographic reviews[4] [13] [11] and an update to 1982, about 780 articles on Azolla have been written since the beginning of the century. It is evident (Figure 1) that the increase has been logrithmic culminating in 1979 the important meeting "Nitrogen and Rice" hold in Los Banos, Philippines. Since then the number of publications has markedly decreased. In this paper I undertake a new appraisal of information and new research highlights on Azolla research.

If one considered the distribution of these articles on Azolla by strain, one notes that there were only 15 publications on the African strains (africana and nilotica), and only three[12] [13] [14] relating to ecological or physiological aspects of A. africana. This scarcity is not correlated with a poor distribution of the strain in Africa. In fact, as the collection of Muséum d'Histoire Naturelle (Paris) can be given an example (Table 1): 66 samples of Azolla presumably A. africana were stored since 1798 whereas there was 72 dried samples of A. caroliniana (the oldest in 1824) and 49 samples of A. filiculoides (the oldest in 1816). These dried samples of A. africana were collected in eighteen countries and from our more recent information A. africana is also present in Burundi, Guinea, Malawi, Mozambique, Liberia, Nigeria, Sierra Leone, Sudan, Tanzania, Zambia and Botswana i.e. in all the tropical and equatorial african countries.

Unfortunately information on their growth conditions and their use are scarce: Decany in 1917 mentioned that in Antanjombala (Madagascar) Azolla, called ramilamina was employed by medicine-men for medicinal purposes. Thus the potentialities of Azolla in Africa remain to be discovered, and, prior to fields experiments, a survey of in situ life conditions must be explored. For this purpose we propose that samples of Azolla collected in Africa should be sent together with soil and water to the ORSTOM station in Dakar for analysis with a standard information form (Table 2) listing important data: the exact place where Azolla was collected, the state, the ecological surrounding, morphological aspects of the strain and its use in the area.

For example consider Senegal where Azolla africana was first mentioned by J. Berhaut (1967)[2] in Nimbato and Casamance, Lebrun (1973)[7] then by Roger and Reynaud (1979)[13] in Bantankutouyel river (Casamance) and later was discovered in Sine Saloum in 1980 by Patrick Jara (ORSTOM, Dakar). In this area an extensive study showed the presence of Azolla in 4 valleys: Koular, Nema, Djikoya and Hamdallai Vilane; the first two only are cultivated (Figure 2). Numerous ethnic groups live in this region and Azolla is called egu-lungate and yangombene in diola; noba, mbaan or gookalek (dress of rice field) in serer; cokok in wolof; koubang or pacoumpacoum in mandingo; thiopotis (weed) in poular, which use Azolla as a vegetable with couscous meal. As far as indigenous populations remember, Azolla was the sole nitrogen input in their rice fields.

A transect analysis in Koular and Nema valleys (Table 3) shown the importance of the fern as an organic and nitrogen manure as the C/N ratio remained constant though C and N concentrations increased 4 fold from the dry cultivated areas (peanut, millet), at the top to the constant overflow bottom. As the cultivation of rice is traditionally carried out by women, agronomic practices remained unknown to African agronomists who did not take this ancillary method into account.

At the workshop on organic matter in West Africa held in Lome, (Togo, 1980) it was reported that the use of Azolla as a N-fertilizer was only effective in Guinea Conakry where the management procedures have ben directly imported from China. Now with the help of West Africa Rice Development Association (WARDA) experiments are in progress in Liberia, Sierra Leone and Senegal.

2. Ecophysiology

In order to measure the inoculation effect of Azolla on rice and on soil, one must first determine the best conditions for growth and the best agronomic practices in a given area. West Africa can be separated into two climatic areas[5]:

- a semi-humid tropical area with a rainy season ranging from five to seven months and conditions similar to Asiatic zones where Azolla grows;
- a dry tropical area with a rainy season ranging from two to five humid months, near the fifteenth parallel.

As Azolla is essential for the expected future nitrogen budget of these lesser developed countries we have to consider its implantations under the extreme climatic conditions prevailing in the latter area: high light intensities (Emax ≃ 90 klux), low temperatures (14°C) and dryness.

2.1 Biological factors influencing N_2-fixation and growth of A. africana

2.1.1 Annual growth curve. During 1981 A. africana was grown in 4 tanks of 1 m^2, collected every 15 days, weighed and reinoculated

at 0.2 kg fw/m^2 . The culture was protected from high light intensities by a mosquito screen. The fastest doubling time was 3 days,
in April-June (Figure 3). At the end of November, the dry season
was well established, night temperatures had decreased and the relative humidity was in the range of 90% to 6 AM to 40% at 2 PM. Then
growth stopped and on the ventral lobe of the fern, micro- and
megasporangae appeared.

2.1.2 Light intensities. Experimental procedure: A tank was divided
into five areas of 25 x 70 cm by floating partitions so that the
culture medium added was identical in each area. Each area was
shaded by a screen allowing transmission of 100%, 60%, 36%, 22%,
7% of incident sunlight, which can reach easily a maximum of 90
klux at 1 p.m. in Dakar.
Each area was inoculated with 40 g FW of A. africana. After 8
and 15 days fresh weight was determined for each area. After 8
days the Azolla morphology was affected by high light intensities:
leaves became red and roots were stunted, under low light intensities
(Emax = 6.3 klux) Azolla was not apparently affected but the productivity was low (Table 4). The ratio clorophyll/carotenoid decreased
with increased incident sunlight. After 15 days the ratio chloro-
phyll/carotenoid did not differ and productivity was proportional
to incident sunlight intensity.
The Acetylene Reducing Activity (ARA), measured at 27°C in different
sunlight intensities, was maximum when Emax. was 60 klux (300 nmoles
$C_2 H_4$ /h/cm^2)$_{1.5}$; higher and lower light intensities (1.7 klux) both
reduced ARA , showing that an optimum energy input is required
to obtain a high level of N_2-fixation.

2.1.3 Temperature. Under optimum light intensity, experiments on
ARA were carried out over the range 25°C-40°C. Temperatures higher
than 35° C increased ARA during the first hour of inoculation but
then inhibited it; at 40°C the ARA was totally inhibited within
6 hours. Optimum ARA of A. africana is in the range of 25-35°C.
In the dry West African climate the growth of A. africana is very
slow, almost stopping between December and April when the temperature
ranges between 15-28°C. In April, when the temperature increased,
the doubling time was reduced to 3 days. This is important as it
is impossible to produce Azolla inocula between December and April,
at these latitudes with the autochthonous strain. However we have
tested A. caroliniana during January 1981 (temperature ranges between
15-28 °C), in a greenhouse and have noticed that its doubling time
was 3 fold less than that of A. africana. The use of this introduced
strain appears promising.

2.1.4 Dryness. During the dry season the relative humidity varied
in 6 hours from 95% to 35%. When the fern is drying up the time
course of ARA falls to zero within 24 h; and remoistening with a
proper growth medium restores neither ARA nor growth [12]. This harmful
effect of desiccation can be slowed down by two processes (summarized
in Table 5):

  1. Adding alginate (0.05% w/v) to the growth medium decreased
     ARA 60% within 24 h; when the fern was remoistened ARA was
     restored to its initial level [12].

2. Stocking in dry condition at 6 °C preserved ARA over more than 48 hours (Figure 4).

2.1.5 Pest control. For the protection of Azolla during the shipment of inocula to the field, we prevent fungus attack [1] (caused mainly by a Myrothecium strain as reported also in Thailand), by spraying cycloheximide (20 ppm) on the fronds (Figure 5). Grazing Lymnaea natalensis was controlled with 2.5 ppm of active carbofuran.

2.1.6 Effect of nature and concentration of combined nitrogen. In the Fleuve region, north of Senegal as much as 120 kg/ha⁻¹ are required for rice cultivation, therefore Azolla has to be used with chemical nitrogen fertilizer.

Azolla was treated with 8.8 and 88 ppm of N-ammonium or N-nitrate. Compared to the N-free medium, nitrate had no effect on growth, 88 ppm of NH_4 -N had a negative effect and 8.8 ppm NH_4-N, a markedly positive effect (Figure 6). High nitrogen concentrations quickly depressed dinitrogen fixation, but in the presence of NO_3-N, fixation was replaced by nitrate assimilation. With 8.8 ppm of combined-N the acetylene reduction of the symbiotic system was depressed to about 40% after 12 days exposure.

More accurate determination of the effect of ammonium-N on Azolla showed that the growth of the fern was stimulated by 44 ppm NH_4 -N, but the ARA was not restored at this concentration, whereas for 1.8 ppm and 8.8 ppm, ARA, which decreased to about one half after 12 days, recovered all its activity after 19 days of incubation (Table 6).

These data show that dinitrogen fixation and combined-N assimilation takes place simultaneously even at 44 ppm concentration of combined nitrogen.[8] Utilization of nitrate seems preferable to the use of ammonium as ammonium competes with dinitrogen for young fronds where they are similarly transported[16]. Our results are in good agreement with these of Yatazawa et al.[17] for A. pinnata var. imbricata is considered as the Asiatic form of A. pinnata var. africana.

3. Plot Assays

3.1 Experiments 1980-81. Azolla pinnata var. africana trials were carried out in Senegal according to the recommendations of the International Network on Soil Fertility and Fertilizer Efficiency in Rice (INSFFER) for 1979. Four plots of 2 square meters, each one containing 950 kg (dw) of sandy soil (N% = 0.140) shrouded in a plastic film to avoid N diffusion, were randomized for each treatment. The trials were carried out at the ORSTOM Station in Dakar (Senegal) for a direct assay during the wet season in 1980 (August to November), then to estimate the residual effect during the following dry season in 1981 (February to June) (Table 7).

During the first assay the climatic conditions were the best for growth of Azolla i.e.: maximum light intensity: 70 klux, maximum relative humidity: 98%, temperature range: 22-37 °C.

To the nine treatments used in other INSFFER trials we added a tenth treatment: Azolla, previously dried at 60 °C and crushed, was incorporated in the soil 10 days before transplanting, at the rate of 60 kg N/ha. In all treatments with N addition, urea was the sole nitrogen source; it was applied three times: before transplanting, 20 days after transplanting and 40 days after transplanting. During the second cultivation cycle, only treatments 2 and 3 received urea-N (Table 8). All the treatments were always provided with tap water.

3.2 Results on the first trial. Inoculation with Azolla always increased the grain and straw yield (Table 7).

Inoculation with Azolla without N fertilizer (treatments 4, 5, 6), increased the grain yield 38-40%, which is similar to the increase resulting from the addition of 30 kg N urea/ha. The increase in the straw yield was higher when Azolla was incorporated (37%) than where it was not (28%). When no N fertilizer was added, the highest yield increase (54%) was obtained with two Azolla inoculations in succession, the first one before, and the second after transplanting (treatment 9). The combination of Azolla inoculation with 30 kg N/ha application (treatments 7, 8) increased the yield more than did 60 kg N-Urea/ha. The yield increases resulting from Azolla inoculation reported here are higher than the average yield increases observed in INSFFER (1980) trials conducted in Asia, due surely in these experiments to the small plot surface and the limited losses of N.

Incorporation of previously dried Azolla (treatment 10) was significantly less favorable than other types of Azolla incorporation. A comparison between treatment 3 (60 kg N-Urea/ha) and treatment 10 (60 kg N/ha as organic N in the form of Azolla powder) shows that the second form of N is less available to rice than the first one.

The growth of Azolla, expressed as N total from Azolla per ha and for each treatment (Table 8), shows that the development of Azolla is significantly better before (treatments 4, 7) than after transplanting (treatments 5, 6, 8). Incorporation always takes place 15 days after inoculation (0.15 kg Azolla fresh weight/square meter). The growth of Azolla was significantly increased with application of urea (treatments 7, 8) at the rate of 30 kg N/ha.

3.3 Residual effect. During the dry season 1981, rice IR 1529 was replaced by KN1H300, a variety assumed to grow better than IR 1529 with cold (15-30 °C) and dry climatic conditions. However, its average yield during the dry season was lower than that of IR 1529 during the dry season. Its straw/grain ratio was also significantly higher (IR 1529: 0.8 and KN1H300: 1.3). With the exception of treatment 9, the low yield of which remains unexplicable, the grain yields and the total N in the soil after two cultivation cycles were higher in treatments with Azolla than in treatments with urea. Thus Azolla inoculation-incorporation treatment bring to the rice, during two cultivation cycles, an equivalent value of 90-120 kg urea-N per hectare. As the total N is higher in Azolla treatment it seems reasonably possible that the residual effect of Azolla inoculation could continue during another rice cultivation cycle.

The most significant event during this cultivation cycle was the good growth of rice in treatment 10: the yield was about the same as in treatments 3, 7 and 8 meaning that, during the first cultivation cycle, the majority of N dried Azolla was not available for rice. We can explain this by the fact that there is, after incorporation of dried Azolla in the wet soil, an aerobic mineralization of only a little fraction of the Azolla material: 20% of total nitrogen was assimilated in 15 days (Figure, 7). But as the 15 first centimeters of soil became quickly anaerobic , denitrification occurred and this mineralized N-Azolla was lost and not assimilated by rice. Anaerobic respiration and formentation transforms Azolla material into fractions easily mineralizable under future aerobic conditions that occur when ploughing for the second rice crop. This absence of effect during the first rice cycle can be limited if Azolla is considered as a green manure, is composted and incorporated just before transplanting.

3.4 Experiments in 1981-82. A second type of trials were carried out according to the recommendations of INSFFER for 1980. The same design as for the 1980-81 plots assays was used with 8 treatments to compare the effect of type of rice transplanted on the Azolla growth and on the rice yield. Results, summarized in table 9 show that:

- there is no effect of the desiccation of rice on the growth of Azolla and on the rice yield;
- Azolla effect on rice yield is more important when it is inoculated after transplanting than before transplanting.

A second rice culture was developed during the wet season 1982 to test the residual effect of Azolla on rice. Caused by some difficulty in germinating rice and in soil preparation, the plots were moistened and dried 3 times before transplanting rice. This unreliable technique gave surprising results: rice yield (grain and straw) in treatments 6, 7 and 8, were Azolla (grown in the previous cultivation cycle was the sole nitrogen input), was significantly lower than treatments with N urea and even than control.

It seems possible that the drying steps involved enhance decomposition of Azolla and that the remoistening steps increased the demineralization process and accordingly increase the C/N ratio causing a soil nitrogen deficiency. On the other hand the formation of phytotoxic compounds can be also considered as an important limiting factor.

Conclusions

- The best growth of Azolla africana occurred during the wet season.
- Nitrate had no effect on Azolla; in the field urea at a rate of 30 kg ha^{-1} had a positive effect.
- Inoculation of fresh Azolla always has a positive effect on the first rice yield.

In regard to the two plot experiments, the effect on a second rice cultivation cycle on the fate of Azolla inoculum incorporated can be positive or negative conditioned mainly by the water control (Figure 8). It appears that prior to proposing a general agronomic procedure on the use of Azolla, studies on its decomposition must be developed under different moistures possibilities with N assays.
The increase of Azolla potential in Africa must involve a large screening of its actual input, the selection of strains adapted

to ecological conditions, the training of technicians and the multi-plication of field trials to perfect the methodology.

Acknowledgements I thank Mr. Pierre Dupont, Michel Zogbi and Patrice Pedurand, for their technical assistance, I am also indebted to Dr. Patrice Jara, for his investigation in Sine Saloum.

References

1 Arunyanart P, Surin A, Rochanahasadin W and Disthaporn S 1982 Rotten disease of Azolla. I.R.R.N. 7, 10-11.
2 Berhaut J 1967 Flore du Sénégal. 2^d edition Clairafrique edts. Dakar, 485.
3 Birch H F 1964 Mineralization of plant nitrogen following alternate wet and dry conditions. Plant and Soil 20, 43-49.
4 Capaya D T 1979 International bibliography on Azolla. Library and Documentation Center, IRRI, Los Banos, Philippines. 66 p.
5 Charreau C 1974 Soils of tropical dry-wet climatic areas of West Africa and their use and management Agronomy Mimeo 74-26 Cornell University, Ithaca.
6 Desvaux C E 1827 Azollaceae Ann. Soc. Limn. Paris 6, 178.
7 Lebrun J P 1973 Enumeration des plantes vasculaires du Sénégal. I.E.M.V.T. (Maisons-Alfort, France) botanical study. No. 2. pp. 209.
8 Liu Chung Chu 1979 Use of Azolla in rice production in China. In: Nitrogen and Rice. IRRI, Los Banos. pp. 375-394.
9 Loyer J Y, Jacq V A and Reynaud P A 1982 Variations physicochmiques dans un sol de riziere inondée et évolutions de la biomasse algale et des populations microbiennes du cycle du soufre. Cahier ORSTOM. Série Biologie, in press.
10 Lumpkin T A, Plucknett D L 1980 Azolla: Botany, physiology and use as a green manure. Economic Botany 34, 111-153.
11 Moore A W, French J E, Dixow H M 1980 Bibliography of Azolla discussion at the workshop on Nitrogen Cycling in South-East Asian Wet Monsoonal Ecosystems - SCOPE/UNEP International Nitrogen Unit - Chieng Mai, Thailand.
12 Reynaud P A 1982 Fixation d'azote chez les cyanobactéries libres ou en symbiose (Azolla); possibilités d'utilisation agronomique en Afrique Tropicale. Bull. Ped. FAO. 47, 63-80.
13 Reynaud P A, Roger P A and Watanabe I 1979 Select bibliography on Azolla with special emphasis on its role in agriculture. Non-Symbiotic Nitrogen Fixation Newl. 7, 7-20.
14 Reynaud P A and Paycheng C 1981 Essai d'inoculation d'Azolla africana dans un milieu peuplé de Lebistes reticulatus. Cah. ORSTOM. sér. Biol. 43, 61-66.
15 Roger P A and Reynaud P A 1979 Premieres données sur l'écologie d'Azolla africana en zone sahélienne (Sénégal). Oecol. Plant. 14, 75-84.
16 Watanabe I, Ke-Zhi Bai, Berja N S, Espinas C R, Ito O and Subudhi B P R 1981 The Azolla-Anabaena complex and its use in rice culture. I.R.P.S. 69, 1-11.
17 Yatazawa M, Tomomatsu N, Hosoda N and Nunome K 1980 Nitrogen fixation in Azolla-Anabaena symbiosis as affected by mineral nutrient status. Soil Sci. Plant Nutr. 26, 415-426.

Table 1: Azolla pinnata var. africana stored in the Museum d'Histoire Naturelle (Paris). Dried samples are classified for each country, the oldest deposited specimen and, if available, some ecological information.

Countries	First Sample	Collected by locality	Dried Samples Stored in M.H.N.	Ecological Information
Angola	1858 Major A. von Mechow	Banza de Libongo	3	1
Benin	1910 A. Chevalier	Porto Novo	1	0
Cameroun	1894 J. Brauw	Yaounde	11	3
Congo	1889 F. Hens	Stanley Pool	6	4
Ivory Coast	1936 P. Chouard	Agnebi	3	3
Gabon	1926 M. Le Testu	Hte Ngounye	2	1
Gambia	1890 J. Brown-Lester		1	0
Ghana	1951 C. D. Adams	Ada	1	0
Guimee C.	1954 R. Schnell	Koba	3	3
Madagascar	1798 G. F. Scott Elliot	Paddies fields	19	12
Mali	1937 M. de Wailly	Gao	2	1
R.C.A.	1903 A. Chevalier	Badi-Chari junction	4	3
Rwanda	1978 J. Raynal	Akagera	1	1
Senegal	1954 R. P. Berthaut	Nema	2	0
Tchad	1964 J. Raynal	Bir Barka	2	2
Togo	1931 G. Mahoux	Nahlkone	1	1
Zaire	1980 H. Vanderyst	Kimpasa	3	2
Zimbabwe	1958 G. F. Cunningham	Victoria falls	1	1

Table 2: Proposition of an _Azolla_ investigation form to collected information of the potentialities of _Azolla_ in Africa.

_____

Collected by: _____
  (name, Christian name)

Employed by: _____
  (name, address, telephone)

              _____

              _____

Local of sampling:
  (Country, region, town, river)

  schematic map:

Date, hour: _____

Water temperature: _____

Water pH: _____

Shade:
  (grass, rice, tree): _____

Associated vegetation: _____

Presence of pests: _____
  (form, species)

Morphological aspect of the fern:

Frond size: _____   root length: _____

Color: _____

Sporangia: _____   mean number for a frond: _____

Sociological importance:

Name of _Azolla_ in vernacular language: _____

People who   used it: _____
  (men, women)

What sort of use: _____
  (nutritional, agronomical, medical)

Other information: _____

Table 3: Analyses of pH, C, N and P_2O_5 on soils transects in Koular and Nema valleys.

Koular Valley	pH water	C %	N %	C/N	P_2O_5 total
Dry top level (peanut)	4.8	2.8	0.28	10.0	0.16
Dry medium level (tomatoes)	4.9	5.2	0.46	11.3	0.18
Wet level (rice)	4.8	8.4	0.94	8.9	0.40
Submerged level (rice)	5.5	14.8	1.22	12.1	0.44
Nema Valley					
Dry top level (millet)	6.4	5.8	0.59	9.8	0.65
Dry medium level (vegetables)	6.6	7.1	0.52	13.7	0.90
Wet level (rice)	6.5	13.2	1.05	12.6	0.46
Submerged level (rice)	5.5	22.2	1.88	11.8	0.49

Table 4: Effect of light intensities on the growth, the leaf coloration and the chlorophyll a content of A. africana.

% sunlight	100	60	36	22	7
Fresh weight after 8 days in g from 40 g of inoculum	115	126	125	148	91
Leaf coloration	green border red	green border yellow	green border yellow	light green	bright green
Chlorophyll a:% dry weight	2.71	–	3.65	–	6.57
Fresh weight after 15 days in g from 40 g of inoculum	264	250	189	215	132
Leaf coloration	pink	pink	light green	light green	bright green
Chlorophyll a:% dry weight	5.83	–	7.1	–	7.8

Table 5: Effect of temperature and alginate (0.05%, 10 mn) on fresh weight and acetylene reducing activity (ARA) of <u>Azolla africana</u>. Each value is the mean of triplicates.

Treatments on fresh Azolla	24 h		48 h	
	% fresh weight	% ARA	% fresh weight	% ARA
R.H.: 98%, 25 °C	94	95	92	90
R.H.: 30%, 25 °C	75	0	47	0
R.H.: 30%, 0.05% alginate, 25 °C	92	45	76	0
R.H.: 98%, 6 °C	97	100	96	88
R.H.: 39%, 6 °C	97	90	94	65

R.H.: relative humidity obtained with K_2SO_4 (98%) and $CaCl_2-6H_2O$ (30-40%).

Table 6: Effect of combined nitrogen $[(NH_4)_2 SO_4)]$ on the growth and acetylene reducing activity of $\underline{A}$. $\underline{africana}$.

Culture solution	% fresh weight			% ARA		
	6 days	12 days	19 days	6 days	12 days	19 days
Control-N	100	100	100	100	100	100
NH_4-N: 1.8 ppm	110	108	100	50	45	95
NH_4-N: 8.8 ppm	112	118	99	45	25	55
NH_4-N: 44 ppm	105	115	50	35	20	5
NH_4-N: 88 ppm	90	80	35	25	12	3
NH_4-N: 175 ppm	85	60	35	20	10	0

Table 7: Effect of A. africana and urea on rice yield and straw conducted as proposed by INSFFER on a wet season (1980) and residual effect on the next dry season in Dakar (Senegal) – rice variety: a: IR 1529, b: KN1H300.

Treatments	Grain a T/ha	Control %	b T/ha	Control %	Straw a T/ha	Control %	b T/ha	Control %
1 Control	3.7	100	2.4	100	3.2	100	3.3	100
2 30 kg urea-N/ha	5.1	138	2.8	117	3.9	122	4.6	139
3 60 kg urea-N/ha	5.5	149	2.7	115	4.4	137	4	121
4 Azolla inc. before transplanting	5.1	138	3.8	161	4.4	137	4.4	133
5 Azolla grown after transp. then inc.	5.2	140	3.5	146	4.6	144	4	121
6 Azolla grown after transp. no inc.	5.1	138	2.7	112	4.1	128	3.8	115
7 30 kg N/ha + Azolla inc. before transp.	5.9	159	3.3	140	4.4	137	3.7	112
8 30 kg N/ha + Azolla after transp. inc.	5.7	154	3	125	4.5	141	3.6	109
9 Azolla grown before and after transp. and inc.	5.7	154	2.4	100	4.3	134	3.3	100
10 60 C dried Azolla, inc. as 60 kg N/ha	4.7	127	3.1	130	3.5	109	4.2	127

inc. = incorporation
transp. = transplanting
a: 1980 wet season
b: residual effect dry season 1981

Table 8: Nitrogen budget calculated by the means of micro-Kjeldahl analysis on two successive rice cultivation cycles (see table 7).

Treatments	N imported as urea kg/ha^{-1}			N from Azolla kg/ha^{-1}	N exported with the yield kg/ha^{-1}			N% in the soil after the two cultures
	a	b	Total	a	a	b	Total	
1 Control	0	0	0	0	45	34	79	0.135
2 30 kg N/ha	30	30	60	0	60	42	102	0.145
3 60 kg N/ha	60	60	120	0	66	40	106	0.149
4 Azolla inc. before transplanting	0	0	0	20	62	51	113	0.160
5 Azolla grown after transp. then inc.	0	0	0	7.5	64	46	110	0.140
6 Azolla grown after transp. no inc.	0	0	0	12.4	61	44	105	0.153
7 30 kg N/ha + Azolla inc. before transp.	30	0	30	34	69	44	113	0.133
8 30 kg N/ha + Azolla after transp. inc.	30	0	30	15	68	40	108	0.164
9 Azolla grown before and after transp. then inc.	0	0	0	35.8	67	34	101	0.149
10 60 C dried Azolla, inc. as 60 kg N/ha	0	0	0	60	55	43	98	0.156

inc. = incorporation

transp. = transplanting

a: rice culture during 1980 wet season

b: residual effect on a second rice culture during 1981 dry season

Table 9: Effect of Azolla and urea on rice yield (variety IR 1529) conducted as proposed by INSFFER on the wet season 1981 and residual effect on the wet season 1982 (after 3 successive desiccations & rewetting) in Dakar (Senegal).

First cycle: wet season 1981	Fresh weight of Azolla incorporated $g\ m^{-2}$	Grain yield $g\ m^{-2}$	Grain yield %	2d cycle: wet season 1982	Grain yield $g\ m^{-2}$	Grain yield %
1 Control no N no Azolla	0	415	100	1 idem	540	100
2 60 kg N.ha^{-1} no Azolla repartition 20 x 20	0	493	119	2 30 kg N.ha^{-1}	500	93
3 60 kg N.ha$^-$ no Azolla repartition 40 x 10	0	451	109	3 60 kg N.ha^{-1}	593	110

4 30 kg N.ha$^-$*
   30 kg N Azolla inc. before
   transplanting, rep.: 20 x 20
	1150	469	113	4 30 kg N.ha^{-1}	592	110

5 As in 4, rep.: 40 x 10*	1292	453	109	5 30 kg N.ha^{-1}	512	95

6 no N, Azolla inc.* before
   transp., reinoculation**
	4375	484	117	6 no N
				   no Azolla
					412	76

7 As in 6 but 40 x 10**	4859	518	125	7 no N no Azolla	410	76

8 no N. inoculation**
   after transpl. 40 x 10
Plot size: 2. s.m.
4 replicates
	2367	509	123	8 no N
				   no Azolla
					490	91

* In 4, 5, 6 and 7 Azolla was inoculated as 0.3 kg fresh weight m^{-2} 18 days before transplanting then incorporated 1 day before.

** In 6, 7 and 8 Azolla was inoculated as 0.3 kg fresh weight m^{-2} after transplanting, incorporated after total cover and reinoculated as 0.3 kg fresh weight m^{-2} 4 times.

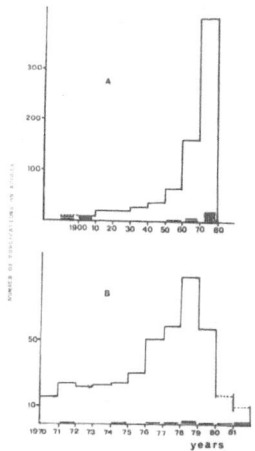

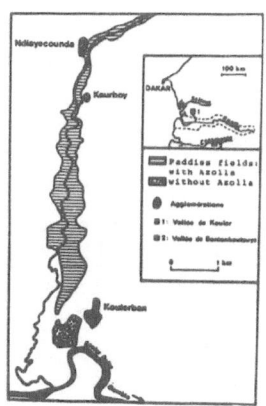

Figure 1.   Diagram of publication on Azolla since the beginning
of the century (A) and for the ten last years (B) with emphasis
on African strains.

Figure 2. Distribution of Azolla pinnata var. africana in South
Senegal: (3) Basse Casamance, (4) Nema Valley; with the map of its
repartition in Koular Valley (Sine Saloum).

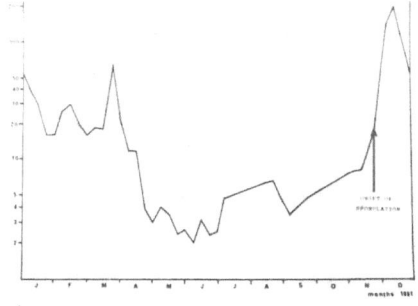

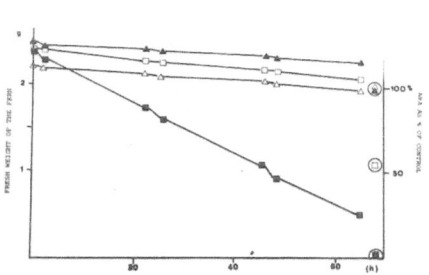

Figure 3. Annual growth curve (1981) of Azolla africana in the
O.R.S.T.O.M. Station (Dakar, Senegal), calculated by the mean of
4 tanks (1 m) collected every 15 days and reinoculated with 0.2
kg fresh weight m ; the growth was under a mosquito screen (Emax:
60 klux).

Figure 4.   Effect of temperature and relative humidity (r.h.) on
the fresh weight and the acetylene reducing activity (ARA) of A.
africana:

▲ —— ▲: 98% r.h., 25 °C   ■ —— ■ : 30% r.h., 25 °C

Δ —— Δ: 98% r.h., 6 °C   ▫ —— □ : 30% r.h., 6 °C

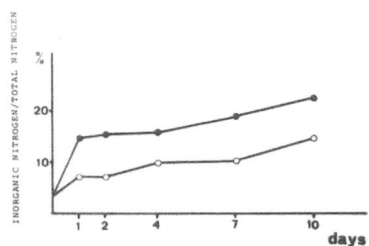

Figure 5. Effect of cycloheximide on the acetylene reducing activity of A. africana stored in moistened conditions.

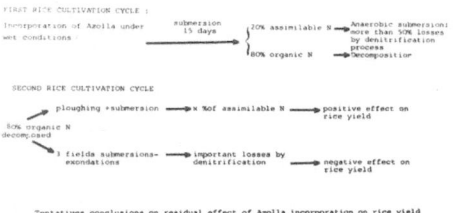

Figure 6. Effect of NO_3-N and of NH_4-N concentrations on the growth and the acetylene reducing activity of A. africana.

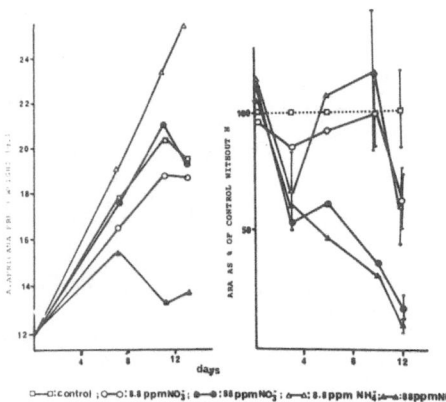

Figure 7. Liberation of inorganic nitrogen in the decomposition of A. africana in moist (• —— •) and dry (o —— o) conditions.

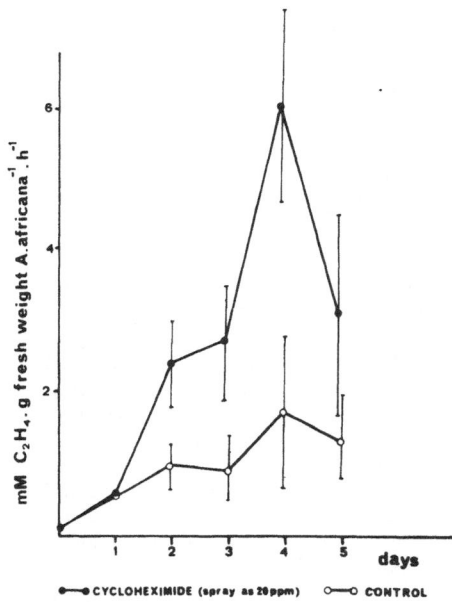

Figure 8. Tentatives conclusions on the residual effect of <u>Azolla</u> incorporation on rice yield in regard to the plots assays conducted in the ORSTOM Station (Dakar, Senegal).

7. A Scanning Electron Microscopic View of Sporulation in Azolla mexicana Presl

H.E. Calvert and G.A. Peters
Charles F. Kettering Research Laboratory
Yellow Springs, Ohio 45387 USA

Key words   Anabaena azollae   Azolla mexicana   Scanning Electron Microscope Sporocarp   Sporulation

The Azolla-Anabaena symbioses are widely recognized as a potentially important alternative nitrogen source for lowland rice. The realization of their full potential remains contingent, however, upon overcoming a variety of physio-ecological and managerial limitations. While ecotype selection programs, the development of improved managerial schemes and the design of specialized machinery all hold promise for improving Azolla utilization, the key to future maximal utilization would seem to lie in understanding the process and control of sporulation and subsequent life cycle development through the haploid gametophyte stage back to the sporophyte. The ability to control the life cycle will facilitate the development of breeding programs and spore inoculants. Few studies have been conducted on the basic reproductive biology of Azolla and to date the factors controlling sporulation remain unknown. As part of an ongoing research program aimed at improving our understanding of the sporulation process and identifying its controlling factors, we have further examined the morphology of sporulation in Azolla mexicana. The result is an improved pictorial record and general confirmation of structures described in earlier literature from other species. Through microdissection sequences the distinctive features within mega- and microsporocarps have been studied and micrographed. We have confirmed the presence of Anabaena in both megasporocarps and microsporocarps. In mixed sporocarp cultures we have followed the dehiscence of the proximal megasporocarp wall to reveal the perispore, rupture of microsporangia which releases the massulae, and attachment of massulae to the megasporocarp by entaglement of the barbed glochidia in thread-like protuberances in the perispore. These cultures have yielded sporophytes. Their emergence and release from the megasporocarp apparatus have been documented pictorially. Taxonomically, our A. mexicana cannot be distinguished readily from A. caroliniana using the details of perispore architecture reported in the literature for the latter species.

Azolla is a free-floating aquatic pteridophyte which maintains a symbiotic relationship with a cyanobacterium named Anabaena azollae Stras. Like all other pteridophytes, Azolla exhibits alternation of generation in its life cycle, producing both a diploid sporophyte and a haploid gametophyte[17]. The Azolla sporophyte produces two different types of spores in distinctive sporocarps. Each spore germinates into a male or female gametophyte which in turn produces male or female gametes. The gametes fuse to form the zygote which develops into the sporophyte.

Sporocarp structure has been described in the literature for all extant species of Azolla except A. mexicana [18 8 3 15 19 5 2 4 7 21 10] [12 16 6 1 11]. Recently we have been able to obtain sporulating samples of A. mexicana and describe here the structure of its sporocarps from examinations with scanning electron microscopy. The A. mexicana examined in this study is from the Gray Lodge, California population obtained from Drs. D.W. Rains and S.N. Talley and determined by Dr. J.W. Hall.

Azolla mexicana bears sporocarps in pairs on the ventral surface of the frond and always at the vertices of frond branches (Figure 1). Megasporocarps are more common as are homocarpous pairs. Microsporocarps are large spherical to oblate bodies, about 1 mm in diameter. Megasporocarps are considerably smaller structures, about 500 µm in length by 300 µm in diameter at their widest point and are fusiform to ovoid in shape. Each sporocarp pair has an associated foliar lamina termed a hood[5] or involucre[10]. The hood, which is only one cell thick, is especially obvious covering nearly mature megasporocarp pairs in A. mexicana.

Each sporocarp is considered a sorus and its covering, the sporocarp wall, to be a modified indusium. In SEM the indusia are seen to be composed of smooth surfaced cells. A whorl of epidermal trichomes is present around the sporocarp receptacle. They are unbranched, terete multicellular hairs, each about 50-75 µm long and 15 µm in diameter. Occasionally Anabaena filaments have been seen around them. Another distinctive external feature of A. mexicana sporocarps is the indusial pore found at the distal end of each sporocarp which bears[14] a certain analogy to the leaf cavity pore of each dorsal leaf lobe. Occasionally Anabaena cells and/or filaments can be seen in and around these pores (see [8], Plate XV).

The microsporocarp contains a mass of spherical microsporangia each approximately 300 µm in diameter (Figure 2). The indusium is two cells thick. Anabaena cells are commonly seen within the microsporocarps. Each microsporangium is borne on a long slender stalk which issues from a central receptacle. The microsporocarp is a fragile structure which easily ruptures releasing its microsporangia. The latter also rupture releasing their contents, the massulae. The surfaces of these alveolate, acellular structures bear elongate barbed processes called glochidia (Figure 3). Normally A. mexicana microsporangia contain four oblate massula in a tetrad-like arrangement. Each massulae is approximately 200 µm in diameter by 100 µm thick, and is pseudocellular in nature. The microspores become entrapped within alveoles of the massulae during their formation in the microsporangium. Each microspore is 10 µm in diameter and bears a characteristic trilete scar. The spore coat lacks any elaborate architecture but does possess a network of low undulating ridges.

Each megasporocarp contains a single megaspore enclosed in a complex structure termed the megaspore apparatus composed of four pseudocellular masses homologous to the microsporocarp massulae. One of the four masses, termed the perispore, encloses the megaspore[12]. The perispore differentiates a spore-covering called the perine and a superstructure which includes the girdle and tricorn-shaped gula (Figure 4). Each of the other three masses is organized into a massulae-like structure called a float, attached to the superstructure of the megaspore.

The megaspore occupies most of the infraspore region of the

megaspore apparatus. It is a single large cell, some 200–300 µm in diameter and bordered by a 5 µm thick spore wall termed the exine or sporoderm[12]. Structural preservation of the cell's cytoplasm is extremely poor. Lucas and Duckett have recently reported the megaspore cytoplasm of A. filiculoides Lam. to be rich in carbohydrate and protein with some lipid inclusions[11]. They have reported that the exine is composed of sporopollenin and carbohydrate.

The surface of the perine is covered with many dimple-like depressions (Figure 3). Closer examination reveals an entwined labyrinth of filamentous processes. The dimples or large depressions in the surface are localized areas devoid of these surface excrescences. Long filamentous hairs which appear to issue from the labyrinthine excrescences extend over the surface of the perine. These hairs, which constitute the megaspore apparatus "capture mechanism"[11], are consistently 1 µm in diameter. Cross fractures of the A. mexicana perine reveal three layers of structure (Figure 5). The innermost layer or endoperine, is 2–3 µm in thickness, has a finely porous nature, and is generally similar in texture to the exine. The outer most layer or exoperine is composed of the labyrinthine filaments already described as the perine surface. The exoperine is 5 µm thick. Between the two is a central layer of tapering columnar excrescences, the mesoperine, which initiates from the inner endoperine and ends with the labyrinthine surface filaments of the exoperine. Overall the perine is approximately 20 µm thick.

Evolutionary modifications of the massular surface processes (glochidia) have produced an effective mechanism, in all Euazolla members, which ensures that the two spore-types and thus the two gametophytic prothalli are in close proximity to each other. Massulae attach to the perine by entanglement of their barbed glochidia in the long hair processes of the perine (Figure 3). Interestingly, the glochidia and the perine hairs are homologous structures, both being elongate surface projections of spore-containing pseudocellular masses formed by tapetal periplasmodium.

A mass of Anabaena cells was always found at the distal end of each megasporocarp under the tip of the indusium (Figure 4, arrow). These cells constitute the inoculum which apparently infects the apical meristem of the nascent sporophyte as it emerges from the megaspore apparatus.

Mixed sporocarp cultures have yielded young sporophytes and we have observed sporophyte release from the megaspore apparatus. The nascent sporophyte emerges out of the distal end of the megaspore apparatus (Figure 6). In the process, the indusium is pushed aside by the sporophyte. Presumably during the time of contact the apical meristem of the young sporophyte is inoculated by the Anabaena mass under the indusium.

The general features of micro- and megasporocarp structure described here for A. mexicana are consistent with the descriptions of other species within the section Euazolla. Perine architecture deserves particular comment since it has been an important character for species delineation. A. microphylla and A. filiculoides have distinctive perine organizations which clearly separate them from each other and from what we describe here for A. mexicana[12,9,16]. The distinction between them and A. caroliniana is likewise clear. However, the distinction between A. caroliniana and A. mexicana is difficult if not impossible to make based on perine structure.

Svenson separated the two by presence or absence of septations in
the glochidia . This feature is now recognized to be variable
and therefore useless as a key character . Svenson did not include
any details of the A. caroliniana perine in his diagnosis because
he found no specimens bearing megasporocarps. The structure of
the A. mexicana megaspore perine described here is generally similar
to the same structure in A. caroliniana based on the descriptions
of Bates and Martin . We should emphasize that we have not as
yet examined sporocarps of our A. caroliniana culture lines because
none have sporulated. But based on present knowledge it seems doubt-
ful that A. caroliniana and A. mexicana can be distinguished based
on the structure of the megaspore apparati.

Acknowledgements

The authors gratefully acknowledge the technical assistance of Sandy
Perkins, Marsha Tootle and Steve Dunbar.

References

1 Bates V M 1980 Re-evaluation of the taxonomy of north american
  Azolla based on megasporocarp morphology. M.S. Thesis, Memphis
  State University, 40 pp.
2 Bonnett A L M 1957 Contribution l'etude des hydropteridees
  III. Recherches sur Azolla filiculoides. Rev. Cytol. Biol. Veg.
  18: 1-85.
3 Campbell D H 1893 On the development of Azolla filiculoides
  Lam. Ann. Bot. (London) 7: 155-187.
4 Demalsy P 1958 Nouvelles recherches sur le sporophyte d'Azolla.
  Cellule 59: 235-268.
5 Duncan R E 1940 The cytology of sporangium development in Azolla
  filiculoides. Bull. Torrey Bot. Club 67: 391-412.
6 Fowler K and Stennett-Willson J 1978 Sporoderm architecture
  in modern Azolla. Fern. Gazette 11: 405-412.
7 Godfrey R K, Reinert G W and Houk R D 1961 Observations on
  microsporocarpic material of Azolla caroliniana. Amer. Fern.
  J. 51: 89-92.
8 Griffith W 1845 On Azolla and Salvinia. Calcutta J. Nat. Hist.
  5: 227-273.
9 Kempf E K 1969 Elektronenmikroskopie der Sporodermis von
  kanozoischen Megasporen der Wasserfarn-Gattung Azolla. Palaont.
  Z. 43: 95-108.
10 Konar R N and Kapoor R K 1974 Embryology of Azolla pinnata.
  Phytopathology 24: 228-261.
11 Lucas R C and Duckett J G 1980 A cytological study of the male
  and female sporocarps of the heterosporous fern Azolla filiculoides
  Lam. New Phytol. 85: 409-418.
12 Martin A R H 1976 Some structure in Azolla megaspores and an
  anomalous form. Rev. Paleobot. Palynol. 21: 141-169.
13 Moore A W 1969 Azolla: biology and agronomic significance.
  Bot. Rev. 35: 17-35.
14 Peters G A and Calvert H E 1982 The Azolla-Anabaena azollae
  Symbioses; in: Algal Symbiosis: A Continuum of Interaction Strategies, L.J. Goff (ed.), Cambridge University Press, NY p. 109-145
15 Pfeiffer W M 1907 Differentiation of sporocarps in Azolla.

Bot. Gaz. 44: 445-454.

16 Pieterse A H, De Lange L and Van Vliet J P 1977 A comparative study of Azolla in the Netherlands. Acta Bot. Neerl. 26: 433-449.

17 Smith G M 1955 Salviniaceae. Cryptogamic Botany. Vol. II. Bryophytes and Pteridophytes. McGraw-Hill, New York, pp. 371-381.

18 Strasburger E 1873 Uber Azolla. Verlag von Ambr. Abel. Jena, Leipzig.

19 Sud S R 1934 A preliminary note on the study of Azolla pinnata R. Br. J. Indian Bot. Soc. 13: 189-196.

20 Svenson H K 1944 The new world species of Azolla. Amer. Fern J. 34: 69-84.

21 Sweet A and Hills L V 1971 A study of Azolla pinnata R. Brown. Amer. Fern J. 61: 1-13.

Fig. 1. Ventral view of an Azolla sporophyte bearing a pair of large microsporocarps (mp), a pair of smaller megasporocarps (Mp) and a heterocarpic sporocarp pair (M & m). vl = leaf ventral lobe, R = root.

Fig. 2. A split-screen, dual magnification micrograph of a microsporocarp with its indusium partially removed to reveal the cluster of spherical microsporangia. The right panel shows an enlarged portion of the inner surface of the indusium. A cluster of Anabaena cells (a) can be seen on the indusium.

Fig. 3. Barb-tipped glochidia (g) of the massulae (ma) entangle in the thread-like surface processes of the megaspore apparatus perine (P).

Fig. 4. The face view of a bisected megaspore apparatus shows the structural arrangement of the megasporocarp including the megaspore (Ms), girdle (gi), gula (gu), floats (f) and indusium (i). Note the mass of Anabaena cells at the distal tip of the sporocarp (arrowhead).

Fig. 5. A cross fracture of the perispore illustrates the architecture of the perine. en = endoperine, me = mesoperine, ex = exoperine.

Fig. 6. A young sporophyte with its single cotyledon (C) emerging from a megaspore apparatus (Ma). Note the displacement of the distal indusium (i).

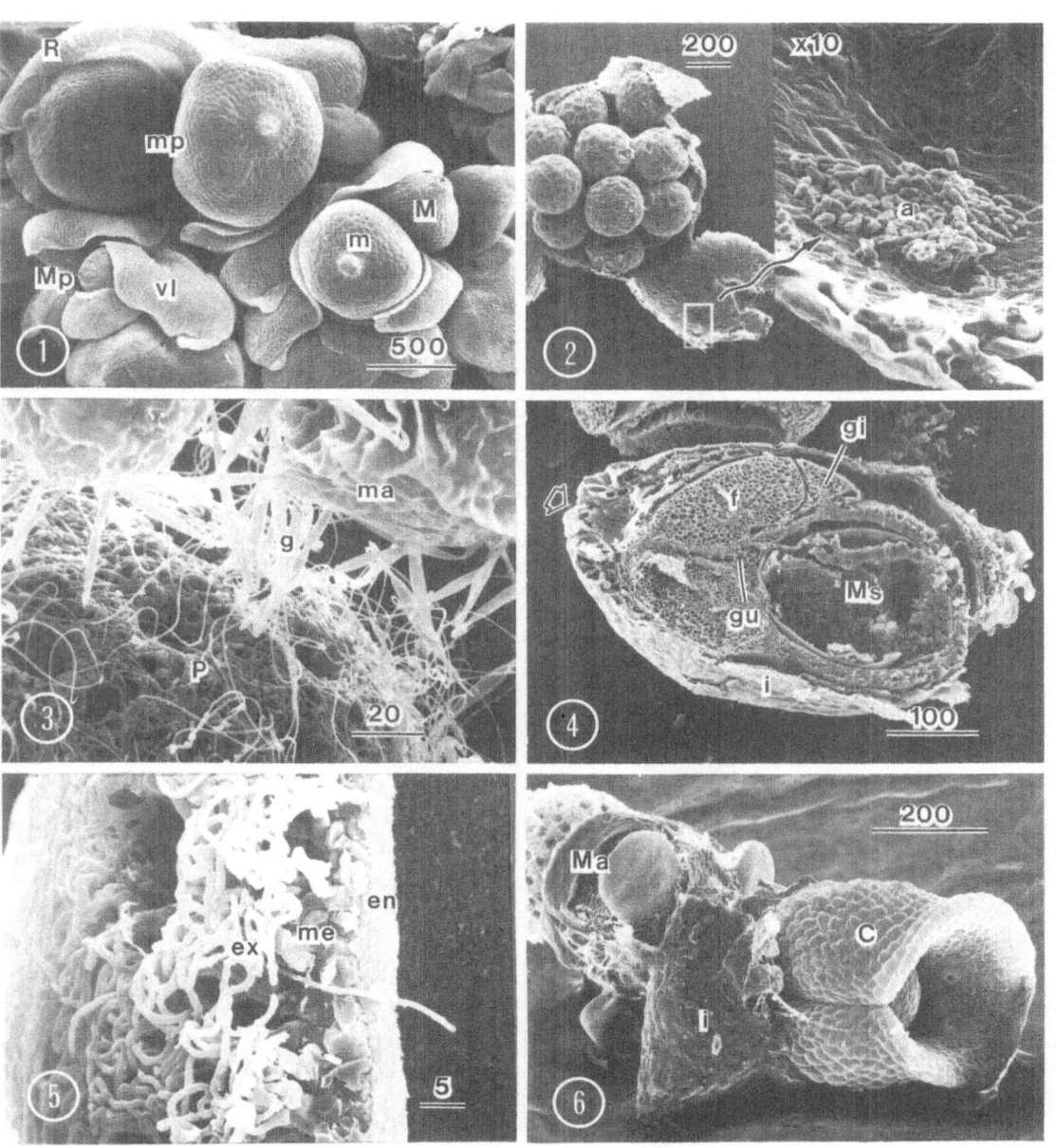

8. Planning Azolla Research for the 1980's

S. N. Talley and E. Lim
Department of Agronomy & Range Science
Plant Growth Laboratory
University of California
Davis, California 95616

Key words Azolla Breeding Mechanization Phosphorus Temperature Tolerance

1. Introduction

During the past five years interest in Azolla as a source of biologically fixed nitrogen for rice has increased dramatically. Azolla cultivation, which developed in conjunction with labor intensive lowland rice cultivation in northern Vietnam and coastal China, is now under study in one or more countries in Africa, Europe, South America, North America, and throughout much of Asia and Southeast Asia. In this paper we review the components of Azolla research we have found useful for evaluating Azolla as a nitrogen source for rice cultivated under mechanization in California, U.S.A. In chronological order these research areas are: defining the limiting conditions for Azolla production within a given region, screening Azolla populations for tolerance to these limiting conditions, optimizing Azolla productivity, and preliminary economic assessment. Concepts related to mechanization of Azolla cultivation and breeding more productive Azolla are also briefly summarized.

2. Basic Components of Field Studies

2.1 Defining Conditions for Azolla Production

Attempts to evaluate Azolla-Anabaena as a nitrogen source for rice at a specific location must first address the climatic limits upon nitrogen-fixation[1,8,13] prevailing when that productivity can enhance rice yield . The climate of the Sacramento Valley (California's principal rice producing region) is the Mediterranean type. A single rice crop with paddy yields approaching 9 t/ha is produced during the warm-dry season between mid May and early September. High yields result from the combination of high insolation, the absence of significant pests or disease, and the use of short stature medium grain rice varieties adapted to heavy doses of inorganic nitrogen fertilizer. All aspects of production are mechanized with field preparation and harvesting taking place, respectively, from early April to mid May and from mid September through October. Field preparation is accomplished after the soil has began to dry at the end of the wet season and harvesting is usually done just prior to the return of precipitation in the fall. A summary of rice production in California is given in a U.S.D.A. publication[11] and by Rutger and Brandon[7].

In temperate climate regions of east central China Azolla is cultivated as a monocrop in May or early June before rice transplanting and as an intercrop with transplanted rice in June[5]. The arid summer climate of California's Sacramento Valley results in high diurnal temperature range. Minimum temperature can drop below 15° C even during the warmest month of July. When this occurs rice yields may decline due to blanking (aborted meiosis of pollen mother cells)[2]. The primary mitigation procedure is early planting – usually no later than mid May. With direct sowing of rice and dry field preparation mandated by the cost and availability of labor and the heavy equipment used in rice cultivation[1] there is little possibility for a spring Azolla monocrop even though nitrogen yields up to 92 kg/ha have been attained by a single Azolla crop at this time[8].

Nitrogen fertilization rates of 135 kg/ha or more as urea, ammonium sulfate, of liquid ammonium hydroxide and stocking with 140 to 225 kg dry wt/ha of previously imbibed rice seed[11] result in a fast growing mat of rice seedlings which cover the water surface within 2 to 4 weeks. Inoculation of Azolla onto these paddies while there is still open water would result in Azolla being moved to one side of the field by wind where it would subsequently impede rice seedling development. If Azolla is inoculated after rice seedlings are erect it would be suppressed within one or two weeks by the developing rice cover. Utilization of Azolla in California rice paddies is, therefore, confined to the period between October or November through March when the fields are fallow.

Preliminary field studies with native Azolla confirm that low temperature results in negligible growth during winter but that during late winter and early spring growth was possible (Figure 1). The slightly alkaline (pH 7.5) flood waters available to rice paddies at this time were also found to be deficient in phosphorus and iron when used as a growth medium for Azolla (Figure 2).

2.2 Screening Azolla for Low Temperature and Low Phosphorus Tolerance

Populations of A. filiculoides outproduce A. mexicana and A. pinnata when tested for low temperature tolerance in outdoor batch culture during the rice fallow-season in California (Figure 3). Intraspecific variation within A. filiculoides respecting low temperature tolerance is several times that observed for A. mexicana from warm-temperate regions (California, U.S.A. and Paraguay). A. pinnata and A. mexicana from tropical regions (Southeast Asia and Guyana, respectively) did not survive the 35 day (20 Feb. – 27 Mar. 1980) outdoor growth trial. Air temperature at the Davis climatological station during this period ranged between 22 and 2°C with the average and standard deviation for the maximum and minimum, respectively, 17 $\pm$ 3 and 5 $\pm$ 2°C[12]. Controlled environment studies utilizing a 15/5°C 12 hr day/night thermoperiod and photoperiod and 1,000 µE m^{-2} sec^{-1} also resulted in mortality of tropical A. pinnata and A. mexicana.

Productivity of A. filiculoides populations is in general agreement with outdoor pot studies. Growth at 250 µE. m^{-2} sec^{-1} minimized the relative differences in productivity within and between Azolla species (Table 1).

Four A. filiculoides populations, two each identified from preliminary screening studies as possessing relatively high (Mono Co., Calif.; Sacramento Co., Calif.) and low (Hawaii Co., Hawaii; Mineral Co., Nevada) tolerance to low temperature were evaluated under regimes of sufficient (3.6 kg P ha^{-1} treatment^{-1}) and insufficient phosphorus (0.9 kg P ha^{-1} treatment^{-1}) in the field between 3 Oct. 1980 and 23 Feb. 1981 (see Table 2 for timing of fertilizer applications and experimental design). During the first 35 days average maximum and minimum air temperature was, respectively 27 and 7°C. Nitrogen accumulation by populations receiving sufficient phosphorus was highest for Sacramento Co. and Mono Co. A. filiculoides. No intra-specific response was noted for populations receiving low phosphorus. Average maximum and minimum temperature decreased, respectively, to 19 and 4°C between day 36 and 46 with freezing of the water surface on the morning of day 42. The Mono Co. population receiving high phosphorus appeared more vigorous (Table 2).

Between 19 Nov. and 5 Feb. the average maximum and minimum temperature was only 13 and 3°C, respectively. Thirteen additional freezing events occurred during this period. The Mono Co. populations receiving 3.6 and 0.9 kg P/ha fertilizations developed multilayered mats containing, respectively, the equivalent of 54 and 68 kg N/ha. Growth was significantly less for Sacramento, Mineral and Hawaii Co. populations which had been disrupted by winds up to 80 km/hr during storms in December. Storm damage in low phosphorus treatments had resulted in termination of net growth. When the experiment was terminated on 23 February, biomass of low phosphorus plots was still at 5 February levels. However, the Mono Co. population receiving high phosphorus now contained the equivalent of 82 kg N/ha. Presence of orange or red-brown macrosporocarps on fronds of all populations except Hawaii suggested nitrogen fixation was at or near termination.

2.3 Optimizing Azolla Production

Growth, nitrogen, and phosphorus content of Azolla used in low temperature-low phosphorus screening studies suggested a mature cover of Azolla could probably be attained during the rice fallow-season using the equivalent of 250 kg fr wt/ha of inoculum and three single superphosphate applications of 1.1, 1.1, and 2.7 kg P/ha after biomass had increased, respectively, 5, 20, and 60-fold. This experiment was carried out between 4 November 1980 and 23 March 1982 using the same A. filiculoides populations listed in Table 2. During early monolayer phases of growth phosphorus uptake was not as high as anticipated - probably because ortho-phosphate uptake by the monolayer fronds was too low to compete successfully with phosphorus fixation by the aerobic paddy soil surface. Nevertheless, the nitrogen yield from the Mono Co. population was over 57 kg/ha after deductions were made for one winter fertilization with KNO$_3$ (1 kg N/ha) and the nitrogen in the 250 kg fr wt/ha inoculum (0.63 kg N/ha). Nitrogen yields from Sacramento Co., Mineral Co., and Hawaii Co. A. filiculoides were, respectively, 85, 75, and 57% of the Mono Co. yield.

Greater protection from wind movement afforded by basal stubble from a previous rice crop in this experiment appears to have been critical to the relatively higher yield obtained from Sacramento Co. and Mineral Co. Azolla relative to the results obtained from the phosphorus-optimization trial. The Hawaii population did poorly in both winter experiments but is equally productive as the remaining three populations on sufficient phosphorus during summer[10]. The greater relative productivity of the Mono Co. versus Hawaii Co. population in winter-fallow fields compared to controlled environment suggests important growth limiting roles for factors which were not simulated in the controlled environment study: turbulence during storms, freezing, day-to-day and diurnal variations in temperature, vapor pressure deficit, and quantum flux. All of these factors except turbulence were fully operational in the outdoor pot study (Figure 3) which is in close agreement with the results of experiments conducted in the field (Table 2).

2.4 Preliminary Economic Assessment

A. filiculoides green manure containing 40-60 kg N/ha and incorporated into soil results in rice yield increases of approximately 1.2 t/ha over unfertilized controls. This is equivalent to the rice yield obtained with approximately 40 kg N/ha as ammonium sulfate (Figure 4). The value of 40 kg N as ammonium sulfate is approximately $25. The cost of phosphorus and other chemicals needed to cultivate a hectare of Azolla in the fallow is approximately $8.50 leaving $16.50 from which the additional and as yet unknown costs to cultivate, collect, disperse, and fertilize Azolla must be deducted to arrive at a final economic value for this green manure crop. With the cost of agricultural labor at approximately $5/hr and aerial application of chemicals at $4/ha, there is doubt that a 250 kg fr wt/ha Azolla inoculum could be cultivated and dispersed onto a rice field for $16.50. Moreover, in the absence of off-season uses for nursery Azolla, like animal fodder, any specialized equipment to produce Azolla could only be used during two or three months of the year due to the highly seasonal nature the Azolla cultivation in rice fields.

3. Recommendations for Future Research

3.1 Application of Azolla Cultivation to a Specific Region

Our experiences in California suggest some general guidelines for conducting applied Azolla studies (Figure 5). Collection and preliminary screening of Azolla will be crucial to the outcome of subsequent research. The higher less variable productivity obtained during winter in fallow-flooded rice fields in the Sacramento Valley with the Mono Co., California A. filiculoides population (which was collected in snow melt waters at 2,500 m elevation) suggest Azolla collections should concentrate upon regions where the climate during the most productive period resembles that of the paddies at the time Azolla production is desired. Preliminary selection of Azolla should be based upon climatic tolerance rather than tolerance to limiting nutrient conditions when a choice has to be made between the two. Tolerance to physical environmental cannot usually be mitigated in the field and, as our field and laboratory studies with A. filiculoides have shown, tolerance to climatic conditions may also confer facultative nutrient-use efficiency. Reliable data upon climatic tolerances of Azolla populations early in the study is so essential we recommend doing all screening studies in the field. If preliminary screening in the field is not possible, outdoor studies using batch culture will at least expose Azolla

hand tools have recently been developed to aid Azolla cultivation
in the P.R.C.[5] but the process remains exceptionally tedious, is
still largely accomplished by hand labor, and is believed to be
economically marginal even in China and Vietnam[5]. We believe mechanization of Azolla production and integration of this process with
mechanized rice production is essential if Azolla is to attain a
widespread and lasting appeal to rice growers.

Mechanized Azolla production will require monocropping and a means
of securing the Azolla cover against turbulence from wind and rain
in large fields. (In California we have kept Azolla in paddies
for up to 143 days during the stormy winter by simply leaving the
stubble from the previous rice crop in place). Incorporation of
Azolla can be accomplished in dry paddies with conventional equipment
or with a wet field cultivator modified for Azolla. The Chinese
have already developed a small scale version of the latter
machine[5]. The additional requirements of mechanized Azolla production
include the machinery to sow, fertilize, spray, harvest, transport,
and disperse Azolla inoculum. This machinery has been conceptualized
in an earlier paper[9] and is summarized here to emphasize that these
machines could be made by modest alterations to existing equipment.
For instance, water tankers and spray rigs could be adapted to fertilize and spray Azolla nurseries - particularly if these nurseries
are long and narrow like agricultural drains. Barges used to harvest
aquatic plants provide a complicated prototype of what would be
needed to harvest Azolla in nurseries. Mechanized production of
vegetable crops relies on large fiberglass bins which ride on flatbed
trailers to transport produce from the field. These same bins could
carry Azolla. Dispersal of Azolla onto paddies could be as complicated as developing equipment to release Azolla from low flying
aircraft or as simple as modifying a manure spreader to handle fresh
Azolla and to move over fallow-flooded fields. Once Azolla is placed
onto fallow-flooded rice fields it can be fertilized with commercial
grade chemicals by low flying aircraft using the same economies
of scale applied to fertilization of the rice crop.

The high cost of agricultural labor, seasonal production schedule,
lengthy period Azolla must remain in winter-fallow fields, and the
inability of existing rice farming machinery to work fields which
are wet combine to render mechanized Azolla production uneconomical
in California. Conversely, the opportunities to develop a mechanized
Azolla technology would be favored in regions with even slightly
reduced wage rates, continuous rice production, relatively uniform
climate favorable to Azolla growth, and where soils permit wet field
cultivation with machinery. Continuous rice production implies
all phases of cultivation are present at a given time and would
optimize the annual return on Azolla nurseries and equipment. Favorable climate might reduce the period for production of a mature
Azolla cover from approximately 140 days reported in this study
to 30 days or less. Climate would also favor Azolla pests but they
could be controlled through the use of moderate amounts (100 g fr
wt/m) of high quality Azolla inoculum, removal of low quality "wild"
or "escaped" Azolla in drainages and ponds near paddies where Azolla
is being cultivated, the timely incorporation of mature Azolla,
and judicious use of pesticides. The ability to cultivate wet rice
fields would streamline the production schedule between Azolla and
rice.

populations to the natural diurnal ranges of temperature and light - environmental parameters which are difficult to duplicate in most controlled environment facilities.

Specific evaluation of factors limiting Azolla productivity will have to take place in the field at the time Azolla production is desired and utilize large enough plots to permit climatic factors such as wind to operate upon the Azolla mat much as it would in a farmer's field. Because the area such experiments require is large we assume only selected populations with promising characteristics will be studied at this level and that some preliminary inputs from bioassays will be used to set the timing and levels of fertilizations and inoculation density.

Field trials will be highly valid predictors of Azolla performance. They must, nevertheless, be validated in controlled environment as this will be the only way laboratory researchers will be able to determine whether or not they can reproduce the level of plant vitality originally observed by field personnel.

Analysis of field experiments to determine factors limiting Azolla productivity will usually provide the data needed to design experiments to optimize N-output from Azolla-Anabaena while minimizing inputs of inorganic nutrients. For instance, biomass, N-content, and P content of samples taken from the 3 Oct. 1980 - 23 Feb. 1981 fallow-season experiment were used to determine inoculation levels, type of field preparation, and nutrient application regime used for the 4 Nov. 1980 - 23 Mar. 1981 phosphorus optimization trial. Other aspects of optimizing Azolla production such as transfer of the Azolla nitrogen to rice in different seasons will be difficult to predict in advance of the actual experiments.

Few economic data have been generated by contemporary Azolla research despite the necessity of such data before applications can be seriuosly contemplated by rice growers. Economic data, when they are presented, are too often like those presented here - preliminary calculations based on only a few experiments which generate a value for Azolla based on equivalence to inorganic nitrogen fertilizers. Beneficial effects of multi-year Azolla cultivation (weed suppression, improvement of soil structure, and subsequent use of "Azolla" nutrients by rice) are ignored[5]. While our calculations excluded weed suppression benefits because fallow-season weeds in California paddies are not the same ones which compete with rice during summer, we generally agree with this criticism. Our data are only intended to provide a conservative preliminary estimate of costs versus returns with Azolla. Professional economic data will require the development of prototype equipment, estimates of the capital and maintenance costs for this equipment in addition to assessment of the land, water, labor, and management costs associated with Azolla production.

3.2 Mechanization of Azolla Cultivation

Where favorable input-output ratios are obtained with Azolla researchers may desire to conduct larger scale experiments. Our field experiences with such larger experiments underscore the prohibitive labor inputs needed to manually cultivate and then incorporate Azolla into rice paddies. These endeavors are estimated to entail 24 to 33 man-days per hectare in the People's Republic of China[5]. Small

3.3 Azolla Breeding

We believe a breeding program is essential to increase Azolla production. We have observed conditions leading to spore production in four Azolla species (Table 3). These data, the significant size differences in mega versus microspores, and the fact that spores survive at least short term (45 day) desiccation without apparent loss of viability, suggests that controlled production and segregation or spores would be possible.

Self-fertilization of A. mexicana and A. filiculoides has been observed by the authors in the field and laboratory. This suggests at least an intraspecific breeding program made possible. Interspecific crosses will be more problematical. For instance, mature spore-bearing A. filiculoides and A. mexicana are occasionally found together in Sacramento Valley agricultural drainages, but we have never found intermediate frond types. Nevertheless, the differential tolerances of A. mexicana and A. filiculoides to temperature and observations of relative desiccation tolerance of some Azolla from Paraguay revealed by our very modest field and laboratory screening trials suggests a breeding program would be useful once there has been significant field collection and screening of Azolla. Without this significant prior investment in population collection from the full range of its native and naturalized habitats and rigorous field screening backed up by controlled environment studies of important characteristics (tolerance to high and/or low temperature, desiccation, salinity, low phosphorus, and pests) the ability to cross different populations or species of Azolla will be of limited value.

Close cooperation between personnel conducting population collection and screening, basic breeding, and progeny screening will, doubtless, present an optimum course for finding Azolla populations with agronomically important characteristics, for pinpointing the bases for these characters, and for producing strains with several combinations of useful traits.

References

1 Berja N and Watanabe I 1981 Response of Azolla species to different temperatures. IRRI Saturday Seminar, March 21.

2 Board J E, Peterson M L and Ng E 1980 Floret sterility in rice in a cool environment. Agron. J. 72: 483-487.

3 Chekiang Academy of Agricultural Sciences, Institute of Soils and Fertilizer, compilers. 1975 Cultivation Propagation and Utilization of Azolla. Agricultural Publishing House, Beijing, P.R.C.

4 Great Britain Meteorological Office. 1958 Tables of Temperature, Relative Humidity and Precipitation for the World. Her Majesty's Stationary Office, London.

5 Lumpkin T A and Plucknett D L 1982 Azolla as a Green Manure: Use and Management in Crop Production. Westview Tropical Agricultural Series, No. 5. Westview Press, Boulder, Colorado.

6 Olsen C 1958 Iron absorption in different plant species as a function of pH value of the solution. Physiol. Plant. 11: 889-905.

7 Rutger J N and Brandon D M 1981 California rice culture. Sci.

Am. 244: 42–51.

8 Talley S N and Rains D W 1980 *Azolla filiculoides* Lam. as a fallow–season green manure for rice in temperate climate. Agron. J. 72: 11–18.

9 Talley S N and Rains D W 1982 Potential mechanization of *Azolla* cultivation in rice fields. In: T. A. Lumpkin and D. L. Plucknett. *Azolla* as a Green Manure: Use and Management in Crop Production. Westview Tropical Agriculture Series, No. 5. Westview Press, Boulder, Colorado. 141–159.

10 Talley S N, Lim E and Rains D W 1981 Applications of *Azolla* in crop production. In: J. M. Lyons, R. C. Valentine, D. A. Phillips, D. W. Rains, and R. C. Huffaker (eds.). Genetic Engineering of Symbiotic Nitrogen Fixation and Conservation of Fixed Nitrogen. Basic Life Sciences, Vol. 17, Plenum Press, New York. 363–384.

11 U S Department of Agriculture 1973 Rice in the United States: Varieties and Production. Agricultural Handbook No. 289, Agricultural Research Service, U.S.D.A., U.S. Government Printing Office, Washington, D. C.

12 U S Department of Commerce 1951–1978, 1980 Climatological Data, California, V55–82, 84. National Atmospheric and Oceanic Data Service, U S Dept Comm, U S Government Printing Office, Washington, D. C.

13 Watanabe I 1978 *Azolla* and its use in lowland rice culture. Tsuchi To Biseibutsu (Soil and Microbe, Japan) 20: 1–10.

Table 1: Productivity of populations representing A. filiculoides, A. mexicana and A. pinnata under 15°/5°C day/night thermoperiod and photoperiod - population codes are described in the caption for Figure 3. Azolla productivity is based on the equivalent of 1 gram dry weight cultivated for seven days on nitrogen free nutrient media. Actual inoculation density was 1.2 g fr wt/pot or 50 g fr wt/m^2. Experiments were conducted in replicates of three and Azolla was harvested six to ten days after inoculation onto the 240 cm^2 surface area of plastic pots containing 2 l of nitrogen free nutrient media previously described[8]. Controlled environment light sources are also described there[8]. Nitrogen determination was by macro-Kjeldahl analysis.

	Quantum flux (μE. m^{-2} . sec^{-1})	
Species/Populations	1,000	250
A. filiculoides		
HA	88 mg N/pot	125 mg N/pot
BX	109	137
SI	132	141
A. mexicana		
G1	0	27
GL	34	46
SB	49	53
A. pinnata		
BN	0	0
BI	0	17
MA	0	16

Table 2: The equivalent content (kg N/ha) of four A. filiculoides population from divergent habitats. Azolla was cultivated in open 15 m^2 (3.3 x 4.6 m) plots between Oct. 3, 1980 and Feb. 23, 1981. Plots were provided the equivalent of 0.9 or 3.6 kg P. ha^{-1}. wk^{-1} as single superphosphate between Oct. 13 and Dec. 8 and again on Jan. 13. Inoculation rate was 25 g fr wt/m^2 and was equivalent to 0.60, 0.68, 0.74 and 0.62 kg N/ha, respectively, for the HA, SI, BX, and WL populations. Results are the average of six subsamples per 15 m^2 plot.

Phosphorus Application Rate (kg P/ha)	Population[1]			
	HA (kg N/ha)	SI (kg N/ha)	BX (kg N/ha)	WL kg N/ha)
35 Days (Oct. 3–Nov. 6, 1980)				
0.9	13.5	14.5	15.0	12.5
3.6	16.8	19.5	22.6	15.1
47 Days (Nov. 7–Nov. 18, 1980)				
0.9	22.2	24.1	22.9	20.6
3.6	29.1	30.2	38.7	34.7
143 Days (Nov. 19, 1980–Feb. 23, 1981)				
0.9	24.8	35.7	50.0	30.4
3.6	43.0	62.3	82.4	65.8

[1]Population codes and collection locations are as follows:
HA: Hawaii Co., Hawaii, USA; collected in a flooded Taro field at low elevation.
SI: Sacramento Co., California, USA; collected in an agricultural drainage at −3 m elevation on Sherman Island.
BX: Mono Co., California, USA; collected in the Owens River at Benton Crossing, elevation 2,500 m.
WL: Mineral Co., Nevada, USA; collected in a seepage spring on the northwest side of the old Walker Lake bed, elevation 1,200 m.

Table 3: A summary of conditions observed to result in spore formation by Azolla. Population codes (when given) are the same as those in Figure 3. Field conditions are those of fallow-flooded fields at the University of California-Davis Rice Research Facility, Yolo Co., California. Temperatures in the Glasshouse did not drop below 20°C but were frequently observed over 40°C in summer.

Species/Population	Observation of Spores	Note and Comments
A. mexicana (GL, SB) Field Glasshouse Controlled Environment	June-October (November) Continuous 20/10-40/30°C, 1000 μE.m^{-2}.sec^{-1} (E) 12 h thermoperiod and photoperiod (T.P.) 25/15-40/30 250 E, 12 hr T.P.	Spores observed in November had formed earlier in the year. Gametophytes and young sporophytes were observed in flooded paddy in May of the following year.
A. mexicana (Paraguay) Field Controlled Environment	July-September (possibly longer) 16/9 100E, 12 hr T.P., 24 hr T.P. 20/15 100 E, 12 hr T.P.	Populations are highly variable respecting spore formation.
A. pinnata Controlled	25/15 250 E, 12 hr T.P. Transfer to 15/5 250 E, 12 hr T.P. (yellow spores). Transfer to glasshouse 32/21 (approx.) 16 hr T, 14 hr P. on P-deficient media (spores ripen on small P-deficient fronds)	Spores would not mature at 15/5 (fronds were slowly dying). Transfer to warmer conditions resulted in growth but spore formation stopped. We used P-deficient media and a warm glasshouse to ripen spores which had formed at 15/5°C and minimize subsequent vegetative growth. Using this procedure fronds were clustered with ripe spores.
A. filiculoides (HA WL SI BX, CM and Tisdale Road, Sutter Co. California) Field Glasshouse	February-November January-May	Spores only formed on multiple layered mats and spore formation ended the period of indeterminant (monolayer) growth. Spore ripening was accompanied by cessation of

Table 3: Continued

Species/ Population	Observation of Spores	Note and Comments
A. filiculoides		nitrogenase activity and followed by frond senescence. Spores over-winter and gametophytes and young sporophytes reappear in April. Spores formed in the glasshouse only when whitewash was removed. Attempts to induce sporulation in controlled environment covered the range of conditions listed for A. mexicana. No spores were observed.
A. caroliniana (Wisconsin)	Field July-August	Spores accompany mat formation but mats are not as thick as in A. filiculoides. Controlled environment studies were as for A. mexicana. No spores were observed.

110

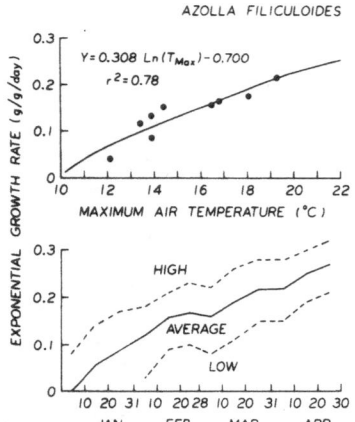

Fig. 1. The potential relationship between temperature and growth rate of A. filiculoides during late winter and early spring in the Sacramento Valley, California. Growth of A. filiculoides was monitored between 4 February and 8 May 1975 in a small agricultural drainage ditch (Tisdale Weir Quadrangle, Lat 39° 1' 27", Long 121° 47' 14"). Maximum temperature data were obtained from the Sutter Basin Reclamation District No. 1660 headquarters 0.7 km northeast of the study area. Exponential growth rate projections were derived from the U.S. Weather Bureau climatological station at Davis, California, between 1967 and 1976 (U S Dept Commerce 1967-1976). Reprinted from Talley et al.[10].

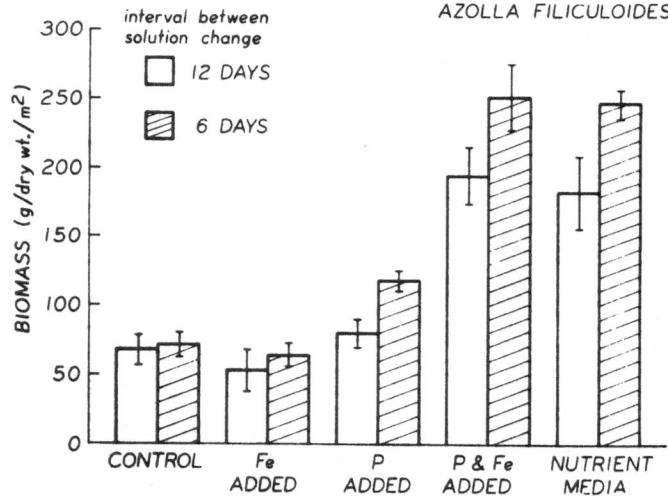

Fig. 2. Biomass of A. filiculoides growing in plastic pots (240 cm[2] surface area) filled with 2 l of Sacramento River water and various nutrient supplements. Phosphorus was added as KH_2PO_4 (12.4 mg P/1). Two and eight tenths mg Fe/1 was added as 0.02 M Fe EDTA[6]. The nitrogen-free nutrient media has been described[8]. The experiment utilized a 1.2 g fr wt/pot inoculation of A. filiculoides which was equivalent to 50 g fr wt, 2.20 g dry wt, 120 mg N and 9 mg P/m[2]. Each treatment was conducted in triplicate outdoors between 11 March and 18 April 1978. Reprinted from Talley et al[10].

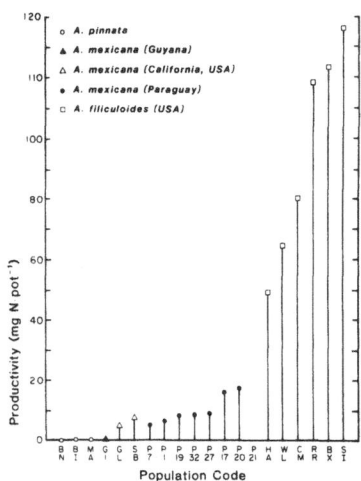

Fig. 3. Kjeldahl nitrogen content of _Azolla_ after cultivating 1.2 g fr wt for 35 days (20 Feb. - 27 Mar. 1980) on the 240 cm^2 surface area of plastic pots. Each pot contained 2 l of nitrogen free nutrient media previously described[8]. Nutrient media was changed weekly. Abbreviations for _Azolla_ populations are: _A. pinnata_ (BN, Banaue, Ifugao, Philippines, IRRI No. 4; BI, Bicol I, St. Domingo, Albay, Philippines, IRRI No. 1; MA, Bumbong Lima, Butterworth, Malaysia, IRRI No. 2), _A. mexicana_ (Gl, Georgetown, Guyana; GL, Butte Co., California, U.S.A., IRRI No. 201; SB, Sutter Co., California, U.S.A.), _A. mexicana_ (tentative) (P7, Filadelfia, Paraguay; Pl, Km 556 Ruta Trans Chaco Paraguay; P19, 32, 27, 17, 20, 21, location uncertain but along Ruta Trans Chaco between Asuncion and the Bolivian border) _A. filiculoides_ (HA, Hawaii Co., Hawaii, U.S.A.; WL, Mineral Co., Nevada, U.S.A.; CM, Cranmore Rd., Sutter Co., California, U.S.A.; RR, Reclamation Rd., Sutter Co., California, U.S.A.; BX, Mono Co., California, U.S.A.; SI, Sacramento, Co., California, U.S.A.).

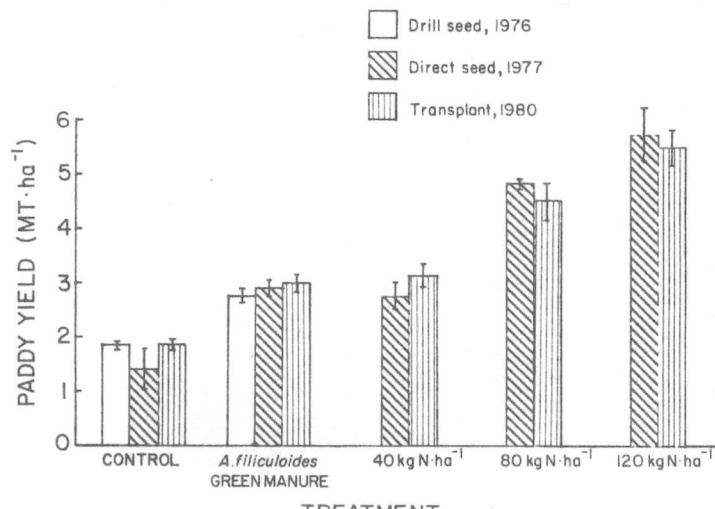

Fig. 4. Effects of A. filiculoides green manure and inorganic nitrogen fertilizer $[(NH_4)_2SO_4]$ on rice yield using drill seed, direct seed, and seedling transplant procedures. Drill seed paddies were planted on 4 May 1976, direct seed paddies were sown on 22 May 1977, and transplanting was done between 20 and 24 May 1980. The nitrogen content of the A. filiculoides green manure was 60, 40 and 55 kg N/ha, respectively during 1976, 1977, and 1980. Adapted from Talley et al.[10]

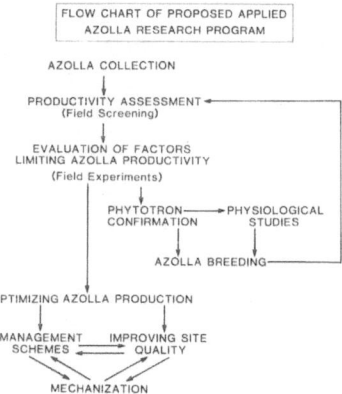

Fig. 5. A suggested sequence for field and laboratory research to determine the productivity and economic viability of Azolla cultivation within a given region.

SECOND SECTION: APPLIED STUDIES

9. Azolla, a water saver in irrigated rice fields?

H.F. Diara
Association por le Développement de la Riziculture en
Afrique de l'Ouest, B.P. 29, Richard Toll, Sénégal

and C. Van Hove
Université Catholique de Louvain, Place Croix du Sud,
4 B-1348 Louvain-la-Neuve, Belgium

Key words Azolla Evapotranspiration Rice

Summary

The influence of Azolla on evapotranspiration has been estimated.
It has been shown that under climatic conditions giving rise to
a 1343 mm evaporation per year Azolla prevents more than 20% of
loss. It is concluded that estimations of the economic impact of
Azolla use have to take in consideration water economy.

Introduction

Since some years research has increased on the potential use of
the aquatic, nitrogen-fixing Azolla-Anabaena azollae association,
principally as a green manure in irrigated rice fields [3]. In some
rice producing areas of tne world, for example in the sahelian
countries, irrigation is expensive, and methods allowing water economy
should be welcome. It was the purpose of the present study to establish the influence of Azolla on evapotranspiration.

Materials and Methods

Two Azolla clones from the catholic University of Louvain collection
were used: Azolla filiculoides Lamarck (strain UCL-9) and Azolla
pinnata R. Brown (strain UCL-7), kindly provided respectively by
Dr. D.F. Toerien, University of Orange State, Republic of South
Africa, and Dr. H.K. Pande, CRRI, Cuttack, India. Plants were grown
in 936 cm^2 plastic boxes (26 x 36 x 12 cm) containing 16 l of nitrate-
free Hoagland-Arnon solution diluted 2/5. Five g Azolla samples
were inoculated in each of seven boxes for each species, while seven
control boxes contain only the solution, to which $CuSO_4$ (10 mg/l)
was added in order to avoid algal development. The 21 boxes were
randomized in a partially controlled, well ventilated, greenhouse
(15 - 33°C; 37 - 72% r.h.; $\pm$ 12 h hemeroperiods).
   Eight days after inoculation Azolla fronds covered practically
all the water surface. The level of culture medium was then adjusted,
every two days as an average, by the addition of measured quantities
of deionized water.

Results and Discussion

Figure I shows the cumulative evaporation and evapotranspiration

curves during a 30 day period, at the end of which <u>Azolla</u> density corresponded to 19.6 t f.wt/ha for strain UCL-7 and 28.5_4 t f.wt/ha for strain UCL-9, values commonly found in the field[2]. From the statistical analysis of the results (ANOVA I) it appears (Table I) that the presence of an <u>Azolla</u> mat on water surfaces significantly reduced water losses. Under climatic conditions giving rise to a 1343 mm evaporation per year, <u>Azolla</u> prevented more than 20% of loss (20.3 and 23.2 according to the species[3]). On a one ha basis this means an economy of 3100 m³ of water which, according to a recent estimation[1] of the cost of irrigation water in the Senegal valley, represents a potential economy of roughly 40 US $.

It is obvious that such an estimation does not take into account many facts, along which the practical impossibility to maintain a permanent <u>Azolla</u> mat, and the interference of rice itself on evapotranspiration, which would tend to decrease the impact of <u>Azolla</u>; on the other hand water loss by evaporation is often much higher under natural conditions than in our experiment; in the Senegal valley for example, it is about 4000 mm/year as measured with a class A evapometric box (H. Van Brandt, personal communication).

The purpose of this short note is therefore only to draw attention to the fact that, in order to estimate the global economic impact of <u>Azolla</u> use in rice fields, data have to be collected on its effect on water loss, which, at least in some regions, may well not be a negligible factor.

References

1 Dachraqui A 1978 Les périmètres villageois de la vallée du fleuve Sénégal. Prix du m³ d'eau d'irrigation. PNUD-FAO-OMVS, fasc. 4, Dt. 230.
2 Lumpkin T A 1977 <u>Azolla</u> in Kwangtung Province, People's Republic of China. Int. Rice Res. Newslet. 2:18.
3 Lumpkin T A and Plucknett D L 1980 <u>Azolla</u>: botany, physiology and use as a green manure. Econ. Bot. 34: 111-153.
4 Talley S N, Talley B J and Rains D W 1977 Nitrogen Fixation by <u>Azolla</u> in Rice Fields. <u>In</u>: Genetic Engineering for Nitrogen Fixation, ed. D. A. Hollaender et al. Plenum Press, N. Y. 538 p.

Table 1: Effect of Azolla on evapotranspiration.

Treatment	Mean final f.wt (g/box)	Mean water loss (ml/box. 30 days)*	Estimated annual water loss (mm)	Estimated annual water economy (% of control)
Control	-	10334	1343	-
A. pinnata (strain UCL-7)	183	8236	1069	20.3
A. filiculoides (strain UCL-9)	267	7941	1033	23.2

* L.S.D. (P=0.01) 681

118

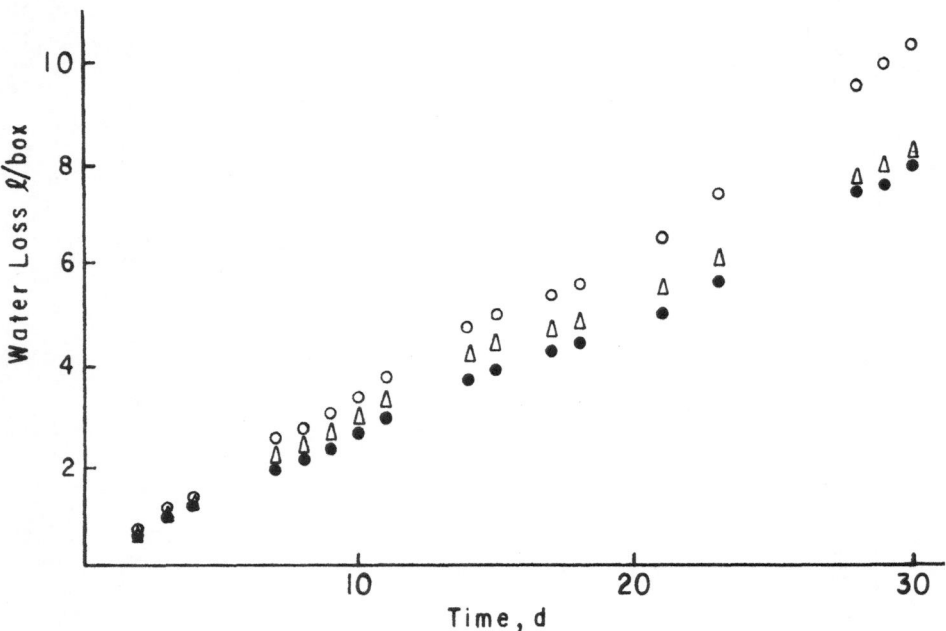

Figure 1.   Cumulated curves of evaporation (O) and evapotranspiration
            as affected by <u>Azolla</u> <u>pinnata</u> (Δ) or <u>Azolla</u> <u>filiculoides</u>
            (●).

10. Azolla - A potential biofertilizer for rice production

S. Kannaiyan, M. Thangaraju and G. Oblisami
Department of Agriculture Microbiology
Tamil Nadu Agricultural University
Coimbatore - 641 003, Tamil Nadu, India

Key words Azolla Eichornia Lemna Manure Rice Yield

Abstract

Multiplication of Azolla under field condition was tried. The application of cattle dung at 10 kg/plot (20 x 2 m) and superphosphate in three split doses at the rate of 100 g/split stimulated the multiplication of Azolla under field conditions. A field trial was also carried out with IR-20 during Samba season - 1981-82. Fertilizer nitrogen at 30 kg N/ha, 60 kg N/ha and 90 kg N/ha with and without Azolla inoculation have been tried. The organic amendments such as Lemna, Eichornia, Azolla and field yard manure were calculated for 30 kg nitrogen equivalent and incorporated. Applications of fertilizer nitrogen along with Azolla inoculation have increased the grain yield considerably. The addition of organic amendments have also increased the grain yield. The increase in straw yield due to Azolla inoculation and addition of organic amendments have been recorded.

Introduction

Azolla is a water fern that fixes atmospheric nitrogen in association with a nitrogen fixing cyanobacterium (blue-green alga) - Anabaena azollae. Nitrogen fixation through the Azolla - Anabaena complex is considered to be a potential biological system for increasing rice yield at a comparatively low cost.[10,19,20] Azolla is extensively used as green manure in Vietnam and China[10,19,20]. The utility of Azolla for rice production in India has also been reviewed[9,15]. Watanabe et al. have established the potential ability of Azolla to fix around 1.1 kg N/ha per day[21] while Singh reported that a layer of Azolla covering a hectare of rice field contained about 10 t of green matter and contributed about 25-30 kg N/ha/y. Azolla can be used effectively as dual cropping with rice to supply nutrients after the formation of a mat either by decomposition after overgrowth or by incorporation into the soil[8]. Azolla inoculated at 0.1 kg/m^2[16], 0.3 kg/m^2[2], 0.2 kg/m^2[7] at 7-10 days after planting was found to establish and cover the area within 15-30 days. Azolla was incorporated in the rice field during the first weeding after the water was drained completely[4]. We report here the multiplication of Azolla under field condition, the effect upon rice yield which we compare to grain yield obtained in response to duckweed (Lemna), waterhyacinth (Eichornia) and farm yard manure supplementation.

Materials and Methods

The field selected for the Azolla nursery was thoroughly prepared and levelled uniformly. The field was divided into one cent plot (20 x 2 m) by providing suitable berms and irrigation channels. Water was maintained at a height of 10 cm and 10 kg of cattle dung was mixed in 20 litres of water and sprinkled on each plot. The inoculum was 8 kg of fresh Azolla. Superphosphate was applied in three split doses at the rate of 100 g/split at 4 days interval as a top dressing fertilizer for Azolla. Furadan granules at the rate of 100 g/plot was applied on 7th day after inoculation. The height of water level was always maintained at 10 cm. The growth and multiplication of Azolla was rapid and in about 10–15 days a thick mat of Azolla was formed on the surface of the water. After 15 days of inoculation the Azolla was harvested and introduced into the main field as a source of primary inoculum. From one harvest 40–55 kg of fresh Azolla was obtained from each plot. Azolla nursery plots may be prepared while the rice nursery is being raised.

A field experiment was conducted during Samba season (September–January – 1981-82) with IR-20 rice variety. Fertilizer nitrogen at 30 kg N/ha, 60 kg N/ha and 90 kg N/ha with and without Azolla inoculation was tried. All the treatments were supplemented with potassium at 50 kg/ha and phosphorus at 50 kg/ha. The green leaf manure such as Lemna, Eichornia and Azolla and the organic manure, farm yard manure (FYM) were incorporated into the plots two days prior to transplanting. These organic sources were calculated to give a 30 kg nitrogen equivalent. Azolla was inoculated as a dual crop at 200 g/m^2 on 7th day after transplanting. The inoculated Azolla multiplied and covered the entire experimental plots within 15 days after inoculation. The first Azolla incorporation was done on twenty-five days after planting. The residual Azolla remaining after the first incorporation multiplied well within the 25 days preceding the second incorporation. Both the grain yield and straw yield were recorded.

Results and Discussion

Phosphorus is known to play an important role in the growth and multiplication of Azolla [4][13]. Since the phosphorus content of the soil or the paddy water was generally too low to meet the growth requirement of Azolla, the addition of phosphorus was neces-sary [7][21]. Phosphorus level at 6 kg/ha has been reported to be optimum for the growth and multiplication of Azolla under field condi-tions [9]. Induction of Azolla growth by split application of phosphorus was reported earlier [8]. In the present investigation the split appli-cation of phosphorus and basal application of cattle dung encouraged the growth of Azolla under field conditions. The addition of cow-dung to the flooded rice soil system has been reported to increase the growth of Azolla as well as nitrogenase activity [13]. The use of cowdung and cattleshed water have shown to increase the growth rate of Azolla during the winter season.

The results of field experiment are presented in Tables I and II. The highest grain yield (5081 kg/ha) was recorded in plots treated with 60 kg N/ha plus Azolla; 90 kg N/ha plus Azolla gave a similar yield (5057 kg/ha). With increases in fertilizer nitrogen along with Azolla inoculation have influenced the grain yield considerably. Among the organic amendments tried, Azolla incorporation gave higher grain yields than by Lemna, Eichornia and FYM incorporation respectively. Clearly, the results indicate that Azolla was effective as a green manure when compared to other treatments. Also Azolla increased the grain and straw yield of rice significantly when Azolla was used as a dual crop. The effective utilization of Azolla as a green manure and in dual cropping have been well documented by several investigators [4] [11] [12] [14] . Govindarajan et al. reported that inoculation of Azolla pinnata at 300 g/m^2 as dual crop with rice and incorporation of Azolla twice showed a significant increase in grain yield of rice which was equivalent to that of 25 kg fertilizer nitrogen per hectare[2]. Srinivasan found that the inoculation of Azolla at 3 t/ha as dual crop has recorded an yield on par with 25 kg fertilizer nitrogen per hectare [18], and Singh reported that a thick layer of Azolla contributed about 30-35 kg N/ha for rice crop per season[16].

The results reported here are in agreement with the observations of many other investigators[4,5,17,21].

References

1 Basavana Gowda R M 1980 Dual cropping and the effect of the combination of inorganic nitrogen requirements and Azolla on the yields of paddy. In: Subject matter training-cum-discussion on Azolla, Mandya, Karnataka, India.
2 Govindarajan K, Kannaiyan S and Ramachandran M 1979 Azolla manuring for rice. Aduthurai Reptr. 3: 89.
3 Kannaiyan S 1979 Nitrogen fixation by Azolla for rice crop. Macao Agri Digest 4: 28-33.
4 Kannaiyan S 1981 Azolla biofertilizer for rice. INSFFER Training Seminar. Int. Rice Res Inst Manila, Philippines. p. 11.
5 Kannaiyan S and Govindarajan K 1980 Effect of Azolla inoculation on rice crop. National workshop on algal systems. Murugappa Chettiar Res Centre, Madras. (Abstr.)
6 Kannaiyan S and Oblisami G 1980 Induction of Azolla growth by split application of phosphorus. 22nd Annu. Microbiol. Conf AMI, Lucknow (Abstr.)
7 Kannaiyan S, Thangaraju M and Oblisami G 1981 Azolla manuring for rice. 22nd Annu. Microbiol. Conf. AMI, Lucknow (Abstr.)
8 Kannaiyan S 1982 Azolla and Rice. In: Training and multiplication and use of Azolla biofertilizer for rice production. Tamil Nadu Agr. University, Coimbatore, Tamil Nadu. p. 35.
9 Kannaiyan S, Thangaraju M and Oblisami G 1982 Studies on the multiplication of Azolla biofertilizer for rice production. National Symp. Biological Nitrogen Fixation, Indian Agricultural Research Institute, New Delhi. p 53.
10 Liu Chung Chu 1979 Use of Azolla in rice production in China. In: Nitrogen and Rice. Int. Rice Res. Inst. Manila, Philippines. p. 375-394.
11 Lumpkin T A and Plucknett D L 1980 Azolla - botany, physiology and use as a green manure. Econ. Bot. 34: 111-153.
12 Moore A 1969 Azolla: Biology and agronomic significance. Bot. Rev. 35: 17-34.
13 Santhanakrishnan P and Oblisami G 1980 Multiplication of Azolla.

In: _Azolla_ a Biofertilizer. Tamil Nadu Agricultural University, Coimbatore, Tamil Nadu. p. 22-26.

14 Singh P K 1977 _Azolla_ plants as fertilizer and feed. Indian Farming. 27: 19-22.

15 Singh P K 1979 Use of _Azolla_ in rice production in Vietnam. In: Nitrogen and Rice. Int. Rice Res. Inst. Manila, Philippines. p. 407-418.

16 Singh P K 1980 _Azolla_ plants as fertilizer and feed. In: Subject matter training-cum discussion on use of _Azolla_. Mandya, Karnataka. p. 1-7.

17 Sree Rangaswamy S R 1980 _Azolla_ biofertilizer for rice crop. In: _Azolla_ a biofertilizer. Tamil Nadu Agricultural University, Coimbatore, Tamil Nadu.

18 Srinivasan M A 1980 _Azolla_ as a biofertilizer for rice corp. In: Subject matter training-cum-discussion on _Azolla_. Mandya, Karnataka.

19 Tran Quat – Thuyet and Deo The Tuan 1973 _Azolla_: a green manure. Vietnamese studies. 38: 119-127.

20 Venkataraman A 1980 Propagation of _Azolla_ in China. In: _Azolla_ a biofertilizer. Tamil Nadu Agricultural University, Coimbatore, Tamil Nadu. p. 1-6.

21 Watanabe I 1977 _Azolla_ utilizationin rice culture. Int. Rice Res. Inst. Newslett. 2: 3.

22 Watanabe I, Espinab C R, Berja N S and Almango B V 1977 _Azolla-Anabaena_ symbiosis. Res. Paper Series 11. Int. Rice Res. Inst. Manila, Philippines. p. 18.

Table I. Comparative effect of _Azolla_ and certain organic manures
on grain yield of rice crop.

Treatments	Grain yield (kg/ha)	Increase in grain yield over control	% increase over control
Uninoculated control	2477	–	–
30 kg N/ha alone	3540	1063	42.91
30 kg N/ha + _Azolla_	4160	1683	67.94
60 kg N/ha alone	4337	1860	75.39
60 kg N/ha + _Azolla_	5081	2604	105.12
90 kg N/ha alone	4533	2056	83.00
90 kg N/ha + _Azolla_	5057	2580	104.15
30 kg N/ha through _Azolla_	4234	1757	70.93
30 kg N/ha through _Lemna_	4057	1580	63.78
30 kg N/ha through FYM	3630	1153	46.54
30 kg N/ha through _Eichornia_	3904	1427	57.61

Grain yield: C.D. = 497.3

Table II. Comparative effect of <u>Azolla</u> and certain organic manures on straw yield of rice.

Treatments	Straw yield (kg/ha)	% increase over control
Uninoculated control	5,680	–
30 kg N/ha alone	7,466	31.44
30 kg N/ha + Azolla	8,573	50.93
60 kg N/ha alone	9,106	60.31
60 kg N/ha + Azolla	9,546	68.06
90 kg N/ha alone	10,040	76.76
90 kg N/ha + Azolla	10,213	79.80
30 kg N/ha through Azolla	7,506	32.14
30 kg N/ha through Lemna	6,933	22.05
30 kg N/ha through FYM	6,346	11.72
30 kg N/ha through Eichornia	6,973	22.76

C.D. = 1680.00

11 Comparative study of six species of Azolla in relation to their utilization as green manure for rice

A. Bozzini
Plant Production and Protection
Division F.A.O.
Rome, Italy

P. De Luca, A. Moretti, S. Sabato and G. Siniscalco Gigliano
Instituto di Botanica
Facolta di Scienze
Universita di Napoli, Italy

Key words Azolla sp. Growth N -fixation Temperate clime

Summary

Seventeen strains comprising six species of Azolla were studied. A. filiculoides was the most suitable for green manure in temperate climates. The endophytes of all species appeared to be identical in morphology although heterocyst frequency varied.

Introduction

Seventeen strains belonging to six species of Azolla coming from representative areas of the world distribution were cultivated in the Botany Institute of the University of Naples (Italy) in order to ascertain their adaptation for their utilization as a green manure in rice cultivation.
  This research concerns the following:
  1. Study of the morphology, taxonomy and reproduction of Azolla and of its symbiont, Anabaena azollae.
  2. Comparative growth of the Azolla species in different environmental conditions (including light, temperature, pH, nutrients).
  3. Selection of strains showing high growth in Naples all year round.
  4. Measurement of nitrogen fixation by Azolla species cultivated under various cultural conditions.
  5. Isolation and culture of A. azollae, and reinfection into alga-free Azolla plants.
  6. Utilization of selected strains as a green manure in experimental rice fields in the Naples Botanical Garden and in cultivated areas.
  In this communication preliminary morphologic, taxonomic and physiologic data on Azolla species cultivated in homogeneous cultural conditions typical of the Mediterranean climate are presented.

Materials and Methods

The list of the strains cultivated at Naples (Italy) is reported in Table 1.
  In the past, many difficulties were encountered in obtaining Azolla

strains since the stress of transport did not permit the survival of these plants. This problem was solved cultivating <u>Azolla</u> on solid inorganic medium KH_2PO_4 5.4 g/l; KCl 14.9 g/l; $MgSO_4.7H_2O$ 19.7 g/l; $CaCl_2.2H_2O$ 29.4 g/l; $FeEDTA$ 0.0385 g/l; A_5 microelement soln 1.0 ml/l; Agar 12%; distilled H_2O 1 l) where <u>Azolla</u> showed good possibility of survival and sometimes even growth (<u>A. filiculoides</u> of Naples). This permitted the transport of the delicate ferns for as long as 12–15 days using readily transportable petri dishes as containers.

At the Naples Botanical Garden <u>Azolla</u> was initially cultivated in liquid inorganic medium (see above). After an acclimation period <u>Azolla</u> was placed in opaque containers (20 cm in diameter x 25 cm high) filled with spring water and soil. The spring water had the following chemical characteristics: hardness $28°$ Fr.; NH_3^- traces; NO_2^- 0.1 mg/l; NO_3^- 0.8 mg/l; Cl^- 105.0 mg/l; SO_4^- 40.0 mg/l.

In order to inhibit undesired algal development, floating plastic grains (perlite) were added into the containers to limit light availability; exceptionaly $CuSO_4$ (10^{-5} M) was added. Cultures were kept in the open air and sheltered from the rain and direct sun rays. Temperature was maintained at 20°C by a thermostatic bath.

<u>Morphology, taxonomy and physiology of</u> <u>Azolla</u> <u>and</u> <u>Anabaena</u> <u>azollae</u>. Morphology of fronds, leaflets, roots and, when possible, of glochidia and massulae were observed as a contribution to the taxonomy of the genus.

The <u>Azolla</u> strains belonged to six species: <u>A. caroliniana</u>, <u>A. filiculoides</u>, <u>A. mexicana</u>, <u>A. microphylla</u>, <u>A. nilotica</u> and <u>A. pinnata</u>. Two strains of <u>A. japonica</u> and <u>A. imbricata</u> could be referred as varieties of <u>A. filiculoides</u> and <u>A. pinnata</u> respectively.

The morphologic distinctive characters of the six species cultivated in homogeneous conditions are as follows:

<u>A. caroliniana</u>: frond subcircular, 1.0–1.5 cm long, smooth, with divaricate branches, dichotomously branched; leaflets 0.5–1.0 mm long, not closely imbricate; roots 1.0–3.0 cm long.

<u>A. filiculoides</u>: frond elongate, 1.5–3.0 cm long, irregularly branched; leaflets oblong to ovate, 1.5 mm long, imbricate, closely appressed; roots 2.5–7.0 cm long; microsporocarps 1 mm in diameter, 35–100 microsporangia, 4–6 massulae, glochidia not septate or with 1 or rarely 2 septae at apex; macrosporocarps 0.1 mm long.

<u>A. mexicana</u>: frond subcircular, 1.5 cm long, dichotomously branched, with compact and thick branches; leaflets 0.7 mm long, lightly detached and raised; roots 1.0–3.0 cm long.

<u>A. microphylla</u>: frond subcircular 0.5–1.0 cm long, dichotomously branched, with branches lightly detached, smooth; leaflets 0.5 cm long; roots 1.5 cm long; massulae, glochidia many-septate, each with about 6 septae.

<u>A. nilotica</u>: frond elongate, 3.5 cm long, triangular, irregularly branched, with very divaricate branches and rachis clearly distinguishable; leaflets 3 mm long, much detached and raised; roots 2.5 cm long.

<u>A. pinnata</u>: frond typically triangular, elongate, 1.0–1.5 cm long, smooth, dichotomously branched, with branches lightly divaricate; leaflets 1.0 mm long, triangular; roots 1.5 cm long, with numerous and very evident lateral rootlets.

With regard to the physiology of Azolla, nitrogen fixation values in one sample of each species were measured. Results are reported in Figure 1 and Table 2. Figure 1 shows that A. filiculoides had higher rates of N_2 than those of the other species. Nitrogen fixation values paralleled growth rates (Table 1) and heterocyst frequency (Table 2). In order to establish the taxonomic identity of the blue-green algae symbiont, the dimension of the vegetative cells and heterocysts and the heterocyst frequency were observed with the light microscope. The shape and dimension of the vegetative cells and heterocysts are similar in the various species, whereas differences among the heterocyst frequencies appear. Many hormogonia but not akinetes were observed. Hormogonia of the various species were morphologically uniform. These results confirm the existence of only one algal species living in symbiosis with all Azolla species [1,2,5,6,7]

Anabaena azollae cultures isolated from A. filiculoides (Naples) by enzymatic digestion were obtained[4]. The isolated symbiont showed the same morphology as Anabaena living in symbiosis and also was able to fix dinitrogen though at very low values (unpublished data). However, the isolated symbiont was kept in viable culture for only 30-40 days.

Growth rates of the Azolla strains. A. filiculoides strains had the maximum growth rates and readily adapted to the Naples cultural conditions (Table 1). Moreover, A. filiculoides adapted more easily to greenhouse conditions. The high growth rate shown by all samples of A. filiculoides is of great interest considering that A. filiculoides strains came from widely separated geographical areas (Italy, U.S.A., Japan). With the exception of A. nilotica, other Azolla species showed lower growth rates.

Of particular interest is the cold resistance of A. filiculoides from Pavia (Italy), collected in a place where, in the winter months, the temperature is often below 0°C.

Conclusion

Among 17 strains of Azolla belonging to the six species of the genus and coming from the whole range of the world distribution, A. filiculoides strains, and in particular the strain of Naples (Italy), showed the most potential for possible utilization in rice cultivation in temperate areas. A. filiculoides, when cultivated in conditions similar to those typical of the Italian springs, showed the highest growth and nitrogen fixation rates. Although not reported in this paper, A. filiculoides of Naples (Italy) grew and fixed dinitrogen in a fairly similar manner during the other seasons of the year.

Among the other species, only A. nilotica presented growth and fixation values fairly near to values of A. filiculoides.

A. caroliniana, A. mexicana, A. microphylla and A. pinnata were not highly adaptable to conditions present in Naples. These last species, therefore, will be studied, in the future, bearing in mind their possible practical utilization in areas with a different climate.

References

1 Fjerdingstad E 1976 Anabaena variabilis status Azollae. Arch

128

Hydrobiol Suppl 49 Algal Studies. 17: 377–381.

2 Geitler L   1925   Cyanophyceae   Geitler L and Pascher A, eds. In: Die Susswasser-Flora. Gustav Fischer, Jena. 329.

3 Lumpkin T A and Plucknett D L   1980   Azolla: Botany, physiology and use as a green manure. Econ Bot 34: 111–153.

4 Newton J W and Herman A I   1979   Isolation of Cyanobacteria from the aquatic fern, Azolla. Arch. Microbiol. 120: 161–165.

5 Shen E Y   1960   Anabaena azollae and its host Azolla pinnata. Taiwania 7: 1–7.

6 Stewart W D P, Fitzgerald G P and Burris R H   1968   Acetylene reduction by nitrogen fixing blue-green algae.   Arch. Microbiol. 62: 336–348.

7 Tilden J   1910   Myxophyceae.   Minnesota Algae.   Report of the Survey, Botanical Series VIII.   Minneapolis, Minnesota. p. 195.

Table 1. Comparative growth of six species of Azolla grown in homogenous conditions of culture*.

Species	Strain	Source	Time for doubling leaf surface (in days)
A. filiculoides	Naples, Italy	(1)	5-7
A. filiculoides	Pavia, Italy	(1)	5.5-7.5
A. filiculoides var. japonica	Japan	(2)	6-8
A. filiculoides	California, U.S.A.	(3)	6-8
A. filiculoides	Nevada, U.S.A.	(3)	6-8
A. filiculoides	Hawaii, U.S.A.	(3)	6-8
A. nilotica	Egypt	(4)	7-9
A. microphylla	unknown origin	(3)	9-11
A. microphylla	unknown origin	(3)	9-11
A. mexicana	Guyana	(3)	9.5-11.5
A. caroliniana	Wisconsin, U.S.A.	(3)	10-12
A. caroliniana	unknown origin	(5)	10.5-12.5
A. mexicana	Mexico	(1)	10.5-12.5
A. pinnata var. imbricata	Japan	(2)	11-13
A. pinnata	Malaysia	(3)	11.5-13.5
A. pinnata	Philippines	(3)	23-29
A. pinnata	Zaire	(5)	28-34

* Culture conditions are reported in the text. Data are referred to period April-June 1982.

(1) Collected by authors. (2) Dr. Ryuso Tanaka, Botanical Institute, Faculty of Science, Hiroshima University, Naka-ku, Hiroshima, Japan. (3) Dr. S. N. Talley, Department of Agronomy & Range Science and the Plant Growth Laboratory, University of California, Davis, California, U.S.A. (4) Dr. G. W. Howard, Biology Department, University of Zambia, Lusaka, Zambia. (5) Plant Production and Protection Division, F.A.O., Rome, Italy.

Table 2. Cell dimensions of Anabaena azollae* and nitrogen fixation of six species of Azolla grown in homogeneous conditions of culture.

Species	Vegetative cells (μm)	Heterocysts (μm)	Heterocystic frequency (%)	Nitrogen fixation ** (nmol ethylene/hr/mg dry wt.)
A. filiculoides (Naples, Italy)	6 x 9	9 x 11	25	2.04
A. nilotica (Egypt)	6 x 8	9 x 12	19	1.46
A. microphylla (unknown origin)	6.5 x 9	9 x 10	17	1.30
A. caroliniana (Wisconsin)	6 x 8	9 x 11	11	0.85
A. mexicana (Mexico)	6 x 8	8 x 10	8	0.79
A. pinnata (Japan)	6 x 8	8 x 10	7	0.38

* Data are referred to period April-June 1982.

** Conditions as in Figure 1.

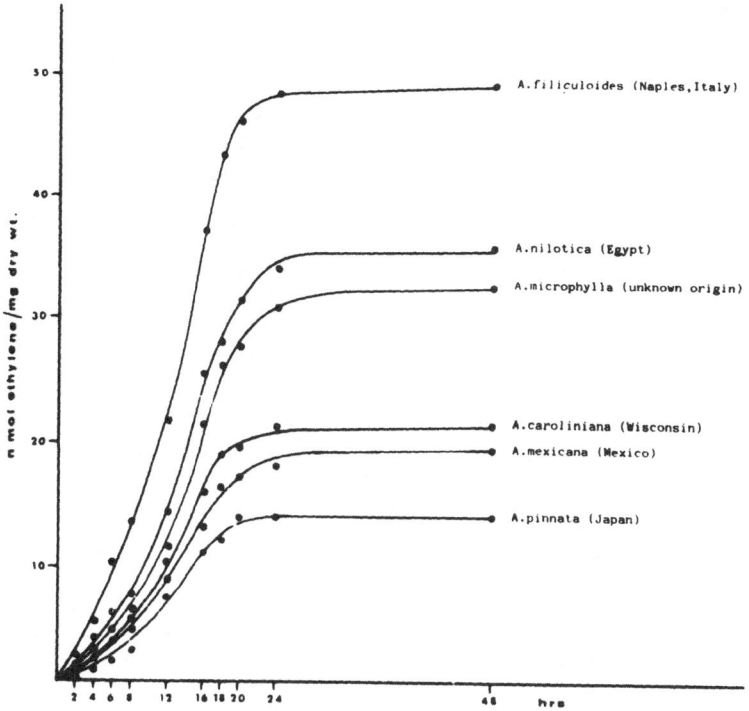

Figure 1. Nitrogen fixation by six species of <u>Azolla</u> grown in homogeneous conditions of culture.

Nitrogen fixation was measured using the acetylene-reduction technique as reported by Stewart et al.[6]. Cultures were incubated at 25 °C and illuminated with fluorescent light of 3,000 lux.
Each point represents the average of three experiments.
Data are referred to the period April-June 1982.

12 Azolla biofertilization to increase rice production with emphasis on dual cropping

P. K. Singh, S. P. Misra and A. L. Singh
Laboratory of Blue-green Algae
Central Rice Research Institute
Cuttack-753006, Orissa, India

Key words   Azolla-Rice dual cropping   Herbicides N_2-fixation
Nitrogen uptake   Rice varieties   Soil fertility

Summary

The response of Azolla dual cropping with three rice varieties Annapurna, Ratna and Jaya of 105, 120 and 140 days duration respectively with and without application of nitrogen fertilizer was studied under field conditions. Inoculation of 500 kg fresh Azolla ha^{-1} after a week of transplanting of rice seedlings, covered the field and formed mat$_1$ in 20 days producing on the average$_1$ biomass of 10.1 to 14.6 t ha$^-$ containing 28.0 to 32.9 kg N ha$^-$ and decomposed after further incubation for a period of 15-20 days. Growth and N-uptake of rice showed that one layer of Azolla was equivalent to 30 kg N ha$^-$ supplied as urea. Growth of Azolla, N-contribution and yield of rice varieties were also similar in both line and randomized planting methods. The same three varieties were also tried with two layers of Azolla, where Azolla basal + dual cropping gave better growth and yield of rice than twice dual cropping.

Growth, nitrogen content of Azolla and rice yield were found to be similar when treated separately with 4.4 kg P kg ha$^-$ and 1000 kg ha^{-1} fresh animal dung. Super phosphate (8.8 kg P) when applied in three splits (10 + 5 + 5) increased Azolla production 50% more than 8.8 kg P applied as basal.

In a separate trial square (20 x 20 cm) and rectangular (40 x 10 cm) rice plant spacings were used where Azolla production, nitrogen contribution and grain yield were similar in both spacings, although east-west direction of planting showed more Azolla production than north-south direction. Five cm of standing water supported better Azolla growth than 10 cm in the planted field.

Azolla growth, N_2-fixation and its response to 85 days duration of rice variety culture-1 was tested in direct seeded crop. The growth and N_2-fixation of basal + dual crop of Azolla was 39.20 t ha^{-1} and 67.58 kg N ha^{-1} which were almost comparable to 130 days rice variety IR-36 under transplanted condition where those values were respectively 46.7 t ha^{-1} and 80.94 kg N ha^{-1}. The grain yield obtained by Azolla basal + dual incorporation was 70.08 and 83% more than the control in transplanted and direct seeded rice respectively whereas Azolla compost, Sesbania green manuring, Eichornia compost and BGA dual increased grain yield to 24.4%, 23.5%, 29.29%, 42.98% and 35.26%, 61.12%, 35.8%, 40.69% over control respectively in transplanted and direct seeded rice. Among the four A. pinnata isolates from Bangkok, Vietnam, Bangladesh and India when tested as basal and dual culture with rice IR-36, Bangkok and Vietnam isolates were superior in growth, N_2-fixation and response to rice crop.

Effect of four herbicides namely, butachlor, benthiocarb, 2,4-DEE and 2,4-DNa was studied at the field recommended doses on fresh weight, dry weight, chlorophyll and N content of Azolla pinnata in unplanted and planted field, where inhibitory effects were observed to be in decreasing order and their effects were nullified after 28 days. Application of field doses of the pesticide carbofuran was found to encourage Azolla growth and N_2-fixation.

Azolla dual cropping was found to be economical. The expenditure involved in production of inoculum + Azolla dual cropping, supplying almost 30 kg N ha^{-1} in rice field was Rupees 38.12 ($4.08), whereas the cost of 30 kg N ha^{-1} as urea was Rupees 140.00 ($14.99). The increase in grain yield at Azolla dual cropping unincorporated and 30 kg N ha^{-1} urea was 0.41 and 0.4 t ha^{-1} over control respectively.

1. Introduction

Azolla is of common occurrence in both tropical and temperate zones and is known as a strong N_2-fixer due to its symbiotic association with the blue-green algae, Anabaena azollae, which remains in its dorsal leaf cavity[4][5][9][10][11]. Various aspects of its importance in the nitrogen economy of rice crop has been studied in Vietnam[19], China[3], Philippines[20] and U.S.A.[18]. It has been introduced in India by Singh[9][10][11]. Besides its use as an organic nitrogen fertilizer[9][10][11], it also checks weed growth to some extent in rice fields[9][10][11]. Since intercropping of Azolla with rice without its incorporation in soil has shown interesting results and this practice is found to be easily adaptable under Indian conditions[11][14][15], it was necessary to intensify the study to obtain information on the effects of different managerial practices on Azolla growth, N_2-fixation and standing rice crop. Phosphorus fertilization has been shown to encourage the growth of Azolla[9][18][19][20], but it was desired to study the resources available to farmers like animal dung which contains 3.5-4.4 kg P 1000 kg^{-1} fresh material. The earlier work conducted at this Institute on Azolla biofertilization has been reviewed[10][11][13][14][17] and an attempt is made in this paper to summarize the findings of the trials conducted during last two years at the Institute Farm.

The present article deals with the Azolla-rice dual cropping with rice varieties considering many parameters: time, effect of line and random planting, effect of square and rectangular plant spacings, the direction of planting, comparison of direct seeded and transplanted rice, water management practice, strains and layers of Azolla, methods and sources of phosphorus application, herbicides, pesticides, decomposition of Azolla in planted field on the Azolla growth, N_2-fixation, N uptake, grain yield of rice and soil fertility.

2. Materials and Methods

2.1 Azolla strains. Azolla pinnata isolated from Vietnam (green isolate), Bangkok, Bangladesh and India (Cuttack) were used in the trials.

2.2 Study area. The field experiments were conducted at Central Rice Research Institute, Cuttack Farm during wet (June to November) and dry (January to June) seasons. This place is situated at 20.05°N latitude, 86°E longitude and 23.48 meter above sea level. The soil is clay loam (32.0% sand, 35.16% clay) rich in K, containing 0.7% organic carbon, 0.006% available P, 0.07% total N with pH 6.5.

2.3 Land preparation. The field was thoroughly ploughed, labelled and plots of 20-25 square meter area were prepared according to the treatments and designs. Randomized block and split plot designs were followed. Rice varieties used in trials were culture one (85 days), Annapurna (105 days), Ratna (120 days),Supriya (125 days), IR-36 (130 days), Jaya (140 days) and CR 1009 (155 days).

2.4 Azolla inoculation. Azolla was grown in the field before as well as after transplanting of rice seedlings. To grow Azolla as basal crop it was inoculated 20 days before transplanting in ploughed, labelled flooded plots with 4.4 kg P ha^{-1} and 2.5 kg ha^{-1} furadan. After 15 days the water of the field was drained and Azolla was incorporated into soil by hand and feet. To grow Azolla as dual culture according to the treatments 100 to 2000 kg ha^{-1}, fresh Azolla was inoculated in the field 7 days after transplanting (DAT), and, after full cover was left for self-decomposition or was incorporated leaving some inoculum for further multiplication for the second Azolla crop (which was again incorporated in some of the trials before the heading stage of rice).

2.5 Transplanting. Seedlings of 20 days old were transplanted at the rate of 2 seedlings hill^{-1} and the damaged plants were replaced within 7 days of planting. The rice plant spacing of 20 x 15 cm, 20 x 20 cm, 40 x 10 cm and random planting were used in the experiments. Direct seeded rice was sown in line at the interval of 30 cm.

2.6 Fertilizer application. Phosphorus was applied in all the treatments at the rate of 4.4 kg P ha^{-1} crop^{-1} of Azolla. This 4.4 kg P was applied into three splits at inoculation time, 5 days and 10 days after inoculation. Higher doses of P were also used in some of the experiments. To protect Azolla from insects 2.5 kg ha^{-1} furadan (75 g a.i. carbofuran) was applied uniformally to all treatments. Urea fertilizer was applied in some of the treatments in three splits (50% as basal + 25% 20 days after transplanting + 25% before heading).

2.7 Azolla biomass determination. In fallow fields the total Azolla was harvested and weighed whereas in planted field Azolla biomass was determined by sampling method for which a square container of 25 x 25 x 25 cm made up of galvanized iron sheet having opening at both ends was placed at random, Azolla present inside was collected and weighed. Azolla collected from different samples of a treatment was mixed thoroughly and 100 g of sample was washed 5-6 times with distilled water, water adhering to Azolla surface was removed with the help of filter paper, and then dried in a oven at 60°C before weighing.

2.8 Herbicides and pesticides. Herbicides, butachlor (a.i. 5%), benthiocarb (a.i. 10%), 2,4-DEE (a.i. 4%) and 2,4-DNa (a.i. 80%) were applied at 0, 7, 14, 21 and 28 days after Azolla inoculation in the field at the field recommended doses of 1.5, 1.5, 1.0 and 0.4 kg a.i. ha^{-1}. Pesticides furadan (3% a.i. as carbofuran) and thimet (10% as phorate) were used at the rate of 2.5 kg ha^{-1} and 1.5 kg ha^{-1} respectively after Azolla inoculation. After incubation of 10 days of herbicides application, fresh weight, dry weight, N content and chlorophyll of Azolla were estimated.

2.9 Grain and straw yield. Grain yield was determined after harvesting the plots leaving two border rows to avoid border effects and the yield was computed to t ha^{-1} at 14% moisture. Moisture content of grain was determined with the help of steinlite moisture meter. Straw was sun dried, weighed and computed to t ha^{-1}.

2.10 Soil and Plant Analysis. Total nitrogen (grain, straw, Azolla and soil) and soil available P were determined by microkjeldahl and Olson methods respectively whereas organic carbon was estimated by Walkley and Black[5] method[1]. The N and P uptake were determined from above grounds parts by computing grain and straw yield with their N and P contents.

3. Results

The results of all recent experiments are presented in Tables 1 to 6. The alloted space for this paper does not permit a detailed presentation of this data, which, however, is thoroughly discussed below.

4. Discussion

High multiplication and fast decomposition rates of Azolla are of special advantage for rice crop[9] [20]. It decomposes in 8-10 days in paddy soil and the rice crop benefitted in 20-30 days[9]. The N content of A. pinnata standing crop has been estimated to be 25 kg N ha^{-1} in paddy[2,20] field as observed in present study. Although the Azolla nitrogen becomes available to the rice crop mainly after its death and decay[4] [19][18] release of NH$_4$-N from growing Azolla is also reported[7]. The release of ammonia in soil is reported to be more rapid from fresh Azolla than from dry Azolla[20] and 68-88% of its release occurs in field within 3-7 weeks of incubation[8][11][16]. In the present study decomposition process was fast in first week and was slower in second or third week, probably due to a swift from aerobic to anaerobic condition of soil. Azolla mineralization in soil is faster than other green manures and its N-release is 41 and 67% in 7 and 35 days in flooded soil[8].

Azolla[19] requires phosphorus fertilizer for its rapid growth. Tran and Dao reported that 1 kg of P$_2$O$_5$ (440 g P) produced a quantity of Azolla equivalent to 2.2 kg of N and recommended 5-10 kg of superphosphate (440-880 g P every 5 days). Singh[9] recommended 2 kg P ha^{-1} week^{-1}. Different methods of phosphorus fertilizer application was tried to increase the Azolla yield and the split application of 8.8 kg P (10 + 5 + 5) was found to be superior as it has increased 50% more Azolla biomass than achieved with 8.8 kg P applied as basal in the present investigation. The split application of 1.1 kg P ha[-1] as superphosphate for 6 times (total 6.6 kg P ha[-1]) increased the Azolla yield more than the basal application and it was equal to 13.2 kg P ha[-1] applied as basal[21]. The 85 and 96% Azolla cover at 21 and 39 days after inoculation in Azolla dual culture with phosphorus and 76% without phosphorus application was reported[20]. The highest growth of A. mexicana was obtained with split application of 15 kg P ha[-1] and its application beyond this level did not enhance growth[6]. In present investigation 9.5-9.7 and 9.0-9.6 t ha[-1] fresh Azolla was observed with phosphorus and animal dung (1000 kg ha[-1]) suggesting that animal dung can substitute phosphorus fertilizer[2,7,10,18].

Azolla suppressed about 50% growth of weeds[4]. Hence, it is likely that herbicides and Azolla need to be used concurrently for the efficient control of weeds. The present investigation on the effect of herbicides indicated that both butachlor and benthiocarb are highly toxic to Azolla and therefore it is suggested that Azolla should not be inoculated in the field during application of butachlor and benthiocarb up to 28 days. However, Azolla can be inoculated after an interval of 28 days, as their effects was nullified. Azolla can also be inoculated along with the herbicide 2,4-DNa causing little damage to Azolla plant. Pesticide carbofuran was found to increase the growth and N₂-fixation of Azolla as reported earlier[10,11].

The Azolla dual cropping (unincorporated) with rice varieties of different durations using low inoculum during dry and wet seasons increased the grain yield and thus it was concluded that this method is also beneficial to rice crop including short duration varieties. Rice yield increased 14, 17, 22 and 40% with cultivation of Azolla[4], A. filiculoides increased 23% yield over control in a trial dual culture with rice[18]. Two successive layers of Azolla incorporated into soil before rice transplanting is reported to supply 50% of N necessary to produce 5 t of rice ha[-1][19]. An increase of 112% and 216% over the control respectively by incorporating one layer and two layers of A. filiculoides was reported[18]. Singh[10,11] obtained 6-39% more yield when A. pinnata was grown with rice but not incorporated and it was 9-41% when incorporated in soil. In the present study increase of 29-48% was observed in grain yield with the incorporation of basal layer and subsequent Azolla dual culture. These values were slightly higher than that obtained with 60 kg N ha[-1] as urea. In general, grain yield and N uptake of Annapurna, which is a short duration variety was less as compared to Ratna and Jaya varieties of longer durations. Singh[14] obtained comparable grain yield with Azolla incorporation as basal plus dual crop to 60 kg N ha[-1]. The incorporation of A. filiculoides in rice paddy soil before transplanting and subsequent dual culture of A. mexicana with rice during dry season was an effective means of increasing N input and rice yield was equivalent to 90 kg N ha[-1] as ammonium sulphate[7]. In the present study Azolla basal + dual cropping twice (incorporation) in square and rectangular spacing resulted more grain yield than 60 kg N ha[-1] as urea and the yield obtained by two layers of Azolla crop (incorporated as dual crop) was slightly less than 60 kg N ha[-1].

The biological nitrogen fixation and application of organic manures are the major sources for maintaining the soil nitrogen pool from which withdrawals are made through N mineralization and subsequent plant uptake. In the present investigation dual culture of <u>Azolla</u> increased the soil nitrogen and organic carbon to 0.082 and 0.86% respectively. Thus soil fertility status was found to be increased due to <u>Azolla</u> dual cropping. The rate of soil N mineralization and the growth and N_2-fixation of <u>Azolla</u> in rice paddy are largely influenced by the ecological factors in soil water ecosystem.

Use of <u>Azolla</u> has been found economical in Vietnam and China although detail information at different locations are lacking. In our study it is found economical than chemical fertilizer but it is suggested to replace furadan with the cheaper pesticide and superphosphate with the locally available resources like animal dung or other wastes. Since, the factors involved in culture and subsequent operational processes varies with time, place and temperature, it is suggested that the economics at different locations must be worked out.

Acknowledgements We are greatful to Dr. H. K. Pandey, Director, Central Rice Research Institute, Cuttack for providing necessary facilities and encouragement.

References

1 Jackson M L 1967 Soil chemical analysis. Prentice Hall Inc. New Delhi, India.
2 Janiya J D and Moody K 1981 Suppression of weeds in transplanted rice with <u>Azolla</u> <u>pinnata</u> R. Br. Paper presented at 12th Annual Conference of Pest Control. Council of Philippines, Laguna, Philippines. pp 1-5.
3 Liu C C 1979 Use of <u>Azolla</u> in rice production in China. <u>In</u>: Nitrogen and Rice. International Rice Research Institute, Philippines. pp 375-394.
4 Moore A W 1969 <u>Azolla</u>: Biology and agronomic significance. Bot. Rev. 35, 17-34.
5 Peters G A and Mayne B C 1974 The <u>Azolla-Anabaena</u> <u>azollae</u> relationship II. Localization of nitrogenase activity as assayed by acetylene reduction. Plant Physiology. 53, 820-824.
6 Rains D W and Talley S N 1978 Use of <u>Azolla</u> as a source of nitrogen for temperate rice culture. <u>In</u>: Proc. Second'review meeting I.N.P.U.T.S. Project. East-West Resource Sys. Inst., Honolulu, Hawaii. May 8-19 pp 167-173.
7 Rains D W and Talley S N 1979 Use of <u>Azolla</u> in North America. <u>In</u>: Nitrogen and Rice. International Rice Research Institute, Philippines. pp 419-431.
8 Saha K C, Panigrahi B C and Singh P K 1982 Blue-green algae or <u>Azolla</u> additions on the nitrogen and phosphorus availability and redox potential of a flooded rice soil. Soil Biol. Biochem. 14, 23-26.
9 Singh P K 197 Multiplication and utilization of fern <u>Azolla</u> containing nitrogen fixing algal symbiont as green manure in rice cultivation. RISO 26, 125-136.
10 Singh P K 1979a Symbiotic algal N_2-fixation and crop productivity. <u>In</u>: Ann. Rev. Plant Sciences. Vol. 1. C P Mallik (ed). Kalyani Publishers. New Delhi, India. pp 37-65.
11 Singh P K 1979b Use of <u>Azolla</u> in rice production in India. <u>In</u>: Nitrogen and Rice. International Rice Research Institute.

138

Philippines. pp 407–418

12 Singh P K 1980 Introduction of green Azolla biofertilizer in India. Curr. Sci. 49, 155–156.

13 Singh P K 1981 Use of Azolla and blue-green algae in rice cultivation in India. In: Associative N_2-fixation Vol. II. P B Vose and A P Ruschel (eds.) CRC Press Boca Raton, Florida. pp 183–195.

14 Singh P K 1982a Azolla as an organic nitrogen fertilizer for medium and lowland rice. In: Review of Soil Research in India. Indian Society of Soil Science and Agricutural Chemistry, Indian Agricultural Research Institute, New Delhi. pp 236–292.

15 Singh P K 1982b Azolla and blue-green algae biofertilizer technology for rice. Indian Farming. 32, 3–8.

16 Singh P K, Panigrahi B C and Satapathy K B 1981 Comparative efficiency of Azolla, blue-green algae and other organic manures in relation to N and P availability in a flooded rice soil. Plant and Soil 62, 35–44.

17 Singh P K, Satapathy K B, Misra S P, Nayak S K and Patra R N 1982 Application of Azolla in rice cultivation. In: Proceeding of the National Symposium on Biological Nitrogen Fixation. Bhabha Atomic Research Centre, Trombay, Bombay. pp 423–450.

18 Talley S N, Talley B J and Rains D W 1977 Nitrogen fixation by Azolla in rice field. In: Genetic Engineering for nitrogen fixation. Alexander Hollander (ed.) Plenum Press, New York, London. pp 259–281.

19 Tran Q T and Dao T T 1973 Azolla: a green compost. Agricultural Problems, 4. Vietnamese studies No. 38, China Books. San Francisco. pp 119–127.

20 Watanabe I, Espinas C R, Berja N S and Alimagno B U 1977 Utilization of Azolla-Anabaena complex as a nitrogen fertilizer for rice. International Rice Research Institute. Res. Paper Series No. 11.

21 Watanabe I, Berja N S and Rosario D C D 1980 Growth of Azolla in paddy field as affected by phosphorus fertilizer. Soil Sci. Plant Nutr. 26, 301–307.

Table 1: Effect of square and rectangular plant spacing on growth and N contribution of Azolla pinnata (Bangkok).

Treatment	Spacing (cm)	Fresh weight (t ha^{-1}) Wet season (1981)	Dry season (1982)	N content (kg ha^{-1}) Wet season (1981)	Dry season (1982)
Azolla basal + 30 kg N ha^{-1}	20 x 20	20.0	20.0	41.5	31.5
Azolla basal + Azolla dual (two layers)	20 x 20	38.1 (a 18.9, b 11.2, c 8.0)	63.1 (a 25.5, b 20.3, c 17.3)	64.4 (a 34.7, b 13.9, c 15.8)	89.3 (a 32.7, b 24.4, c 32.2)
Azolla dual (two layers)	20 x 20	26.0 (b 14.0, c 12.0)	39.3 (b 18.5, c 20.8)	43.8 (b 17.7, c 26.1)	60.7 (b 27.3, c 33.4)
Azolla dual + 30 kg N ha^{-1}	40 x 10	20.0	20.0	34.5	28.3
Azolla basal + Azolla dual (two layers)	40 x 10	42.6 (a 19.8, b 12.7, c 10.1)	64.2 (a 25.8, b 20.8, c 17.6)	75.8 (a 41.8, b 15.7, c 18.3)	94.0 (a 35.9, b 24.8, c 33.3)
Azolla dual (two layers)	40 x 10	29.3 (b 15.0, c 14.3)	40.6 (b 19.3, c 21.6)	42.9 (b 18.2, c 24.7)	58.8 (b 24.4, c 34.4)

Figures in parentheses are separate biomass and N content of Azolla as
a: Azolla basal; b: Azolla dual 1st layer; c: Azolla dual 2nd layer

Table 2: Growth and N content of Azolla pinnata (Vietnam) (two layers) in the planted rice field.

Treatment	1981 (wet season)						1982 (dry season)					
	Fresh weight (t ha^{-1})			N content (kg ha^{-1})			Fresh weight (t ha^{-1})			N content (kg ha^{-1})		
	Annapurna	Ratna	Jaya	Annapurna	Ratna	Jaya	Annapurna	Ratna	Jaya	Annapurna	Ratna	Jaya
Basal + Dual cropping	22.1	23.0	24.0	58.4	58.5	60.1	21.4	22.4	22.5	58.5	59.6	58.5
Azolla dual incorporated	12.6	12.6	12.6	30.0	30.0	30.0	12.0	12.0	12.0	30.0	30.0	30.0
Azolla dual cropping	9.5	10.4	11.4	28.4	28.5	30.1	9.4	10.4	10.5	28.5	29.6	28.5
Azolla dual cropping (twice incorporated)	18.4	19.0	19.9	52.6	52.3	52.6	17.0	17.2	19.9	52.9	55.9	55.0
1st layer	10.6	11.1	12.0	29.3	30.0	29.2	9.9	10.2	12.0	28.8	30.4	29.3
2nd layer	7.8	7.9	7.9	23.3	22.3	23.4	7.0	7.0	7.9	24.1	25.5	25.7
Azolla dual cropping (un-incorporated)	10.2	10.4	10.9	29.3	29.4	29.7	10.0	11.0	11.9	29.2	30.6	27.5

Table 3: Effect of herbicides on Azolla pinnata (Vietnam) (dual cropping) and on grain yield of rice variety Ratna (growth and N content of Azolla were determined after 12 days of incubation).

Treatment	1981 (wet season)		1982 (dry season)		Grain yield (t ha^{-1})	
	Fresh weight (t ha^{-1})	N content (kg ha^{-1})	Fresh weight (t ha^{-1})	N content (kg ha^{-1})	1981 (wet season)	1982 (dry season)
Control (No herbicide, no Azolla)	-	-	-	-	1.5	2.5
Azolla (No herbicide)	8.7	27.9	8.3	27.6	1.9 (26.7)	3.2 (28.0)
2,4-DEE + Azolla*	4.7	14.3	4.6	14.1	1.7 (13.3)	2.5 (4.0)
Butachlor + Azolla*	0.5	1.4	0.5	1.5	1.5 (0.0)	2.5 (0.0)
Benthiocarb + Azolla*	2.0	5.6	2.2	6.3	1.5 (0.0)	2.5 (0.0)
2,4-DNa + Azolla*	8.6	28.5	9.2	28.1	2.0 (33.3)	3.2 (28.0)
2,4-DEE + Azolla**	5.0	14.5	5.0	12.7	1.7 (13.3)	2.6 (4.0)
Butachlor + Azolla**	0.7	2.6	0.7	2.6	1.5 (0.0)	2.5 (0.0)
Benthiocarb + Azolla**	2.4	6.0	2.7	2.6	1.5 (0.0)	2.5 (0.0)
2,4-DNa + Azolla**	9.0	28.0	8.9	28.8	2.0 (33.3)	3.2 (28.0)
C.D. at 5%					0.07	0.09
C.D. at 1%					0.09	0.13

Figures in parentheses indicate % increase over control.

* Azolla and herbicides applied together.

** Azolla inoculated seven days after herbicide application.

Table 4: Effect of two layers of *Azolla pinnata* (Vietnam) and inorganic fertilizer on grain yield and N uptake of rice varieties.

	1981 (wet season)						1982 (dry season)					
	Grain yield ($t\ ha^{-1}$)			N uptake ($kg\ ha^{-1}$)			Grain yield ($t\ ha^{-1}$)			N uptake ($kg\ ha^{-1}$)		
Treatment	Anna-purna	Ratna	Jaya	Anna-purna	Ratna	Jaya	Anna-purna	Ratna	Jaya	Anna-purna	Ratna	Jaya
Control	2.4	2.4	2.4	35.9	35.5	35.5	2.6	2.5	2.5	37.7	37.2	36.9
Azolla basal + dual (un-inoculated)	3.0 (25.0)	3.1 (29.2)	3.2 (33.3)	50.6 (41.0)	50.6 (42.6)	49.6 (39.8)	3.9 (26.9)	3.2 (28.0)	3.2 (28.0)	56.2 (49.0)	55.1 (48.2)	54.7 (48.2)
Azolla dual twice inoculated	2.9 (20.8)	2.9 (20.8)	3.1 (29.2)	48.7 (35.7)	49.1 (38.4)	53.0 (49.4)	3.1 (19.2)	3.0 (20.0)	3.1 (24.0)	52.7 (29.8)	51.7 (39.0)	52.9 (43.4)
Azolla dual (uninoculated)	2.6 (8.3)	2.6 (8.3)	2.7 (12.5)	41.4 (15.4)	42.7 (20.3)	44.2 (24.5)	2.8 (7.7)	2.8 (12.0)	2.8 (12.0)	45.5 (20.7)	45.4 (22.1)	45.6 (23.6)
60 kg N ha^{-1}	2.9 (20.8)	3.1 (29.2)	3.1 (29.2)	49.4 (37.5)	52.3 (47.4)	52.9 (49.1)	3.1 (19.2)	3.2 (28.0)	3.1 (24.0)	52.9 (40.3)	55.5 (49.2)	53.3 (44.4)
C.D. to compare the means		5%	1%		5%	1%		5%	1%		5%	1%
of the varieties		N.S.	N.S.		0.86	1.99		N.S.	N.S.		N.S.	N.S.
of two treatments between varieties at the same level of treatments		0.05	0.07		0.73	0.99		0.05	0.07		7.6	1.0
		0.08	0.1		1.3	1.7		N.S.	N.S.		1.3	1.8
of two treatments at the same or different levels of varieties		0.1	0.1		1.4	2.1		N.S.	N.S.		2.1	2.9

Figures in parentheses indicate difference from control in percent

N.S. = Not significant

Table 5: Effect of Azolla pinnata (Bangkok) and chemical nitrogen fertilizer on grain yield and N uptake of rice variety Ratna.

Treatment	Spacing (cm)	N applied (kg N ha⁻¹)		Grain yield (t ha⁻¹)		N uptake (kg N ha⁻¹)	
		Wet	Dry	Wet	Dry	Wet	Dry
Control	20 x 20	0.0	0.0	1.86	2.89	30.3	39.1
60 kg N ha⁻¹	20 x 20	60.0	60.0	2.98 (60.0)	3.88 (34.0)	48.4 (61.4)	64.6 (65.3)
Azolla basal + 30 kg N ha⁻¹	20 x 20	71.5	61.6	2.90 (56.0)	4.03 (40.0)	53.2 (75.7)	64.2 (64.3)
Azolla dual + Azolla dual twice	20 x 20	64.6	89.3	2.82 (52.0)	4.57 (58.0)	47.4 (56.7)	75.5 (93.1)
Azolla dual twice	20 x 20	43.8	60.7	2.71 (46.0)	3.66 (27.0)	43.9 (44.9)	55.0 (40.6)
Control	40 x 10	0.0	0.0	1.98	2.78	32.9	38.6
60 kg N ha⁻¹	40 x 10	60.0	60.0	2.83 (43.0)	3.86 (39.0)	46.9 (42.7)	58.7 (52.7)
Azolla basal + 30 kg N ha⁻¹	40 x 10	74.5	58.3	3.26 (65.0)	4.1 (47.0)	55.4 (68.6)	66.6 (72.8)
Azolla basal + Azolla dual twice	40 x 10	75.8	94.0	2.71 (37.0)	4.57 (64.0)	50.3 (53.0)	87.0 (125.6)
Azolla dual twice	40 x 10	42.9	58.8	2.72 (37.0)	3.68 (31.0)	44.9 (36.5)	60.4 (56.5)
C.D. to compare difference between treatment means	5%			0.235	0.316	3.49	8.0
	1%			0.331	0.446	4.71	10.8
Spacing means				N.S.	N.S.	N.S.	N.S.
Interaction means				N.S.	N.S.	N.S.	N.S.

Figures in parentheses indicate % increase over control.
N.S. = Not significant

Table 6: Economics of dual cropping of Azolla.

Cost of inoculum production

Plot area – 80 m² divided into 8 sub-plots of 8 m² each
Inoculum = 3 kg fresh Azolla/plot
Azolla produced after a week = 82 kg fresh/64 m²

	Rs.	$
Superphosphate 360 g	0.36	0.04
Furadan 15 g	0.30	0.03
One laborer for Azolla inoculation (35 min)	1.20	0.13
One laborer for superphosphate and furadan application (15 min)		
One laborer for irrigation (twice) (30 min)		
Pump expenditure for irrigation	1.00	0.11
Two laborers for harvesting Azolla (30 min)	0.50	0.05
Labor for field preparation & maintenance for one crop	1.00	0.11
Total cost	4.36	0.47
Cost of 500 kg inoculum production for one hectare	34.62	3.71
Labor cost for inoculation in rice field	3.50	0.37
Total expenditure for one hectare	38.12	4.08
*Furadan 2.5 kg	50.00*	5.35*
*Superphosphate 65 kg	65.00*	6.96*
Cost of urea equivalent to 30 kg N ha^{-1} (65 kg)	140.00	14.99
Net gain	101.88	10.91

* Since superphosphate (4.40 kg P ha^{-1}) and pesticide furadan (2.5 kg ha^{-1}) are also applied for rice crops, their cost is not included.

13. Manuring of rice crop with Azolla

B. Roy
Senior Scientist Rice Agronomy
Agricultural Research Institute
Mithapur, Patna, India

Key words   Azolla   N uptake   Rice

Summary

Field experiments conducted for 3 years (1978-1980) in wet (Kharif) seasons under neutral soil conditions for determining the efficacy of Azolla as compared to urea revealed that (i) response of rice to N was 24, 15 and 10 kg N/ha at 25, 50 and 75 kg N/ha applied as urea, respectively, (ii) response to 6 t/ha of Azolla alone (incorporated) was equivalent to 36 kg N/ha, (iii) response to Azolla (inoculation, growing in situ and incorporated) was equivalent to 24 kg N/ha.

1. Introduction

The energy crisis and consequent increased cost of chemical N fertilizer, the widening gap between the indigenous supply and demand, coupled with the low purchasing power of the cultivator, have imposed serious limitations on rice production. Increasing the dependence of biological nitrogen fixation is the alternative being suggested and emphasized for certain parts of the world.

   Azolla pinnata R. Brown, an aquatic fern containing Anabaena azollae in symbiotic association, is commonly found in India in ponds, ditches and channels containing stagnant water[7] and the nitrogen fixed by the blue-green algae is available to rice plant[1,4,7]. Azolla has been used for the fertilization of rice fields in Central China, North Vietnam, Indonesia and Thailand for a long time[5,9]. To assess the effectivity of this fern on nitrogen economy in rice cultivation in the neutral soil of Patna, an experiment was conducted during 1978 to 1980 under the All India Coordinated Rice Improvement Programme. The results of these experiments are presented here.

2. Materials and Methods

The experiment was conducted at the Agricultural Research Institute Farm, Patna, Bihar (at 25.3 N, 85.15 E and 57.8 m above sea level). The experimental soil was heavy clay in texture and medium in fertility (organic 0.60%, total N 0.076%, available P and K 16 and 147 kg/ha respectively, CEC 16 meq/100 gm of soil) and neutral in reaction (pH 7.1 in 1:2.5 soil water ratio). A detailed list of the treatments are given in Table 1. The experiment was conducted in a randomized block design with four replications in 20 m^2 plots. In the first year Azolla was obtained from the Central Rice Research Institute, Cuttack which was then propagated at the Institute Farm at Patna for trial in subsequent years. The growth of Azolla from middle of June to October was very rapid but low ambient temperature greatly reduced the rate from November to February. During summer months its survival in open fields was difficult, and hence it was shaded in order to maintain satisfactory growth. The inorganic nitrogen was applied as urea in three splits. A uniform basal dose of 25.8 kgP and 33.2 kgK was applied through single superphosphate and muriate of potash respectively. Application of 6 tonnes of fresh Azolla was incorporated manually before planting after draining out the water. After incorporation it was again multiplied and incorporated. In other treatments where Azolla was only inoculated @ 1.0 tonnes/ha it was applied 3-4 days after planting in 8-10 cm water. It was further multiplied within 10-12 days which was incorporated into the soil. In all the three Kharif crop seasons, a medium duration variety 'Sita' was grown. The crop was transplanted in the second week of July and harvested in the last week of October. Soil and plant samples were taken after the harvest of the crop and were analyzed for organic carbon by the Walkly and Black method and nitrogen by a modified Kjeldahl method as described by Jackson (1967).

3. Results and Discussion

Application of Azolla influenced all the chemical properties of the soil considerably (Table 1).

3.1 Organic carbon: Application of six tonnes of fresh Azolla or Azolla with inorganic nitrogen increased the organic carbon status of the soil (0.66%), which was higher than the treatment receiving 75 kg N/ha through inorganic forms (0.63%). Even the Azolla inoculated plot had as much organic C (0.64%) as was recorded in plots receiving 75 kg N/ha.

3.2 Total nitrogen content in soil: The total nitrogen content in soil was also high in the treatment receiving six tonnes of Azolla alone or 6 tonnes of Azolla with 25 kg N/ha, followed by the treatment where 50 kg and 75 kg N/ha as urea.

3.3 Mineralized nitrogen: Mineralized nitrogen content was the highest in the treatment where 6 tonnes of Azolla was applied with 25 kg of inorganic nitrogen, followed by 6 tonnes of Azolla incorporated plots and Azolla inoculated plots.

3.4 Nitrogen uptake: Total nitrogen uptake was the highest in the treatment where 75 kg inorganic nitrogen was applied, followed by the treatment where only Azolla was inoculated, incorporated (6 t/ha) and incorporated @ 6 t/ha with 25 kg N/ha; the differences between these treatments were, however, not significant. A similar trend of variations in the nitrogen contents in straw and grain of rice was also observed.

3.5 Number of panicles/m^2 and sterility percentage in spikelets: In all the three years the highest number of panicles/m^2 was found in rice crop receiving 75 kg N/ha in inorganic form, followed by treatment receiving 50 kg N/ha and 6 tonnes of fresh Azolla (incorporated at planting). Six tonnes of Azolla incorporated at planting and 25 kg inorganic N/ha also gave comparable number of panicles/m^2 (Table 2). The sterility percentage was the lowest (10%) where rice crop was fertilized with 75 kg N/ha followed by the treatment receiving 50 kg N/ha (12%) as compared to 13.6% in the treatment where rice was fertilized with Azolla (6 t/ha incorporated).

During Kharif 1978 the response to N was 15, 14 and 15 kg grain/kg N added as urea at 25, 50 and 75 kg N/ha, respectively (Table 2). Response to six tonnes fresh Azolla alone (incorporated) was equivalent to 11.2 kg N/ha. Azolla inoculation and growing in situ with rice was equivalent to 13.9 kg N/ha. During Kharif 1979 the response to N was 31, 24 and 21 kg/kg of N added as urea at 25, 50 and 75 kg N/ha respectively. Response of six tonnes of fresh Azolla was equivalent to that of 30 kg N/ha. Azolla inoculation in situ and growing with rice was equivalent to 29.8 kg N/ha. During Kharif 1980 the response to N was 26, 14 and 19 kg/kg N applied as urea at 25, 50 and 75 kg N/ha respectively. Response of six tonnes of Azolla alone was equivalent to that of 39.7 kg N/ha. Even Azolla inoculation alone gave an additional grain yield equivalent to 13.2 kg N/ha. Year to year variations were mainly due to seasonal variations in the growth and yield of rice and the mineralization of nitrogen in the soil.

Combined results of all the three crop seasons indicated that the response of N at different nitrogen levels were 24, 15 and 10 kg grain per kg of N applied as urea at 25, 50 and 75 kg N/ha respectively. Response to six tonnes of Azolla alone accounted for an additional grain yield equivalent to that of 36 kg N/ha. Even Azolla inoculation without other supplements gave an additional grain yield equivalent to 24 kg N/ha.

Acknowledgement   The author is thankful to the Project Director, All India Coordinated Rice Improvement Project   Hyderabad for providing materials for the trial. He is also thankful to Sri P.N. Ghosh, Regional Director and Dr. R. C. Chaudhary, Chief Scientist Rice for providing facilities for experiment and to Dr. B. N. Chatterjee, Professor of Agronomy Bidhan Chandra Krishi Vishwa Vidyalaya, Kalyani for correcting the manuscript.

References

1 Becking J H   1975   Nitrogen fixation in some neutral ecosystems in Indonesia. In: Symbiotic Nitrogen Fixation in Plants. Nutman P C, ed. Cambridge University Press. 539-550.
2 Becking J H   1979   Environmental requirements of Azolla for use in topical rice production. In: Nitrogen and Rice Inter. Rice Research Inst. Manila, Philippines. 345-374.
3 Jackson M L   1967   Soil Chemical Analysis. Parentice-Hall of India Pvt. Ltd. 498.
4 Jain H K   1978   Algal Technology for Rice Research. Bulletin No. 9. Indian Agricultural Research Institute, New Delhi, India.
5 Liu Chung Chu   1979   Use of Azolla in rice production in China. In: Nitrogen and Rice. Int. Rice Res. Inst. Manila, Philippines. 375-393.
6 Singh P K   1977   Multiplication and utilization of the fern Azolla containing its nitrogen fixing symbiont as green manure in rice cultivation. IL Riso 26: 125-136.

7 Singh P K 1977 Effect of Azolla on the yield of paddy with and without the application of N fertilizer. Curr. Sci. 46: 642–644.

8 Singh P K 1979 Use of Azolla in rice production in India. In: Nitrogen and Rice. Inter. Rice Res. Inst. Manila, Philippines. 407–418.

9 Tuan D T and Thuyet T O 1979 Use of Azolla in rice production in Vietnam. In: Nitrogen and Rice. Inter. Rice Res. Inst. Manila, Philippines. 395–405.

Table 1: Effect of _Azolla_ application on chemical properties of soil and nitrogen uptake in rice.

Treatments	Organic Carbon %	Total N %	Mineralized N kg/ha	N uptake		
				Total (grain + straw) kg/ha	Grain kg/ha	Straw kg/ha
Control (no supplements)	0.60	0.079	239	55.30	29.22	26.08
25 kg N/ha as urea	0.61	0.092	240	75.86	38.20	37.66
50 kg N/ha as urea	0.61	0.094	241	79.59	39.93	39.66
75 kg N/ha as urea	0.63	0.094	244	82.35	41.71	40.54
Azolla 6t/ha (incorporated)	0.66	0.096	250	81.22	42.76	38.46
Azolla inoculated*	0.64	0.088	248	82.04	43.04	39.00
25 kg N/+ 6t/ha of _Azolla_ (incorporated)	0.66	0.096	251	81.63	49.32	32.31
Mean	0.63	0.091	245	76.83	40.59	36.24
C.D. at 5%	N.S.	0.004	N.S.	4.93	2.63	3.65
C.V. %	5.2	4.5	3.5	4.5	4.5	7.1

*Growing _in situ_ and incorporated.

Table 2: Grain yield and yield contributory characters or rice in response to various supplements.

Treatments	Panicles/m^2				Mean Sterility (%)	Grain yield (kg/ha)			
	1978	1979	1980	Mean		1978	1979	1980	Mean
Control (no supplements)	305	205	275	295	18.8	3856	2477	3409	3246
25 kg N/ha as urea in 2 splits	396	227	355	326	15.4	5236	3255	4053	4181
50 kg N/ha as urea in 2 splits	356	241	342	313	12.1	5479	3665	4129	4358
75 kg N/ha as urea in 2 splits	395	277	373	348	10.2	5959	4059	4848	4852
Azolla 6t/ha (incorporated)	362	274	301	312	13.6	5025	3407	4432	4455
Azolla inoculated*	380	246	247	291	14.1	5068	3405	3750	4161
25 kg N + 6t/ha of Azolla (incorporated)	387	257	289	311	12.6	5494	3945	4072	4378
Mean	386	248	296	310	–	5160	3576	3981	4179
C.D. at 5%	72	18.1	11.7	–	–	305	233	452	–
C.V. %	9.5	8.7	6.8	–	–	8.2	5.4	11.0	–

* Growing in situ followed by incorporation.

14. Phosphorus removal by <u>Azolla</u> <u>caroliniana</u> cultured in nutrient enriched waters

K.R. Reddy and W.F. DeBusk, respectively
Associate Professor and Biologist
Agricultural Research and Education Center
Institute of Food and Agricultural Sciences
University of Florida
Sanford, Florida 32771

Key words   <u>Azolla</u> <u>caroliniana</u>   Nutrient enriched waters   Phosphorus

Summary

Phosphorus removal by <u>Azolla</u> <u>caroliniana</u> cultured in nutrient enriched waters was evaluated under varying environmental and cultural conditions. <u>Azolla</u> growth rates were found to be influenced by plant density, temperature, solar radiation and nutrient composition of the culture medium. Biomass yields were found to be highest during first harvest (5.6 to 7.9 g dry wt m^{-2} day^{-1}) and yields decreased significantly during subsequent harvests, when plants were cultured without replenishing the depleted nutrients. Phosphorus removal in <u>Azolla</u> systems was found to be due to plant uptake and chemical precipitation. Phosphorus removal due to plant uptake was found to be in the range of 16.7 to 92.1 mg P m^{-2} day^{-1}. Estimated N_2 fixation rates were found to be in the range of 201 to 275 mg N m^{-2} day^{-1}.

1. Introduction

Aquatic plants have received increased attention in recent years as a possible system for reducing nutrient levels of wastewaters. Under natural conditions, freshwater aquatic plants can be seen in dense stands in polluted waters producing large amounts of biomass that has no value and constitutes an aesthetic and environmental problem. Efficient management of these plants either in natural or artificial systems containing nutrient enriched waters can improve water quality while the byproduct, the plant biomass, can be potentially used for feed, fiber, and conversion to gaseous fuels. Although many aquatic plants were found to be very efficient in reducing nitrogen (N) levels of wastewaters, their phosphorus (P) removal potential is limited, thus the effluent leaving these treatment systems usually contains unacceptable levels of P[1,2].

   The purpose of this investigation was to evaluate the P removal potential of <u>Azolla</u> <u>caroliniana</u> cultured in effluent containing very little or no available N. <u>Azolla</u> was selected because of its ability to grow in waters containing very little or no N, and its symbiotic relationship with the N-fixing blue-green alga.

2. Materials and Methods

<u>Azolla</u> used in this study was collected from the field reservoirs

located in Zellwood, Florida. Initially, plants were cultured in
300 l outdoor tanks containing minimal amounts of nutrients minus
N. Phosphorus concentration of the water medium was < 0.5 µg
ml^{-1}. The description of the experiments is given as follows:

Experiment I:

This study was designed to evaluate the effect of seasonal variability
on P removal from the water. Azolla was cultured in 1000 l concetre
vaults having a surface area of 1.7 m^2. Nutrient medium containing
no N, 3.1 µg P ml^{-1}, 23.5 µg K ml^{-1}, 20.0 µg Ca ml^{-1}, 5.0 µg Mg
ml^{-1}, 2.0 µg Fe-EDTA ml^{-1}, and minor elements were used in the study.
Once a week water in the vaults was replaced with fresh nutrient
medium. Azolla, 200 g m^{-2} fresh weight, was added to each of two
vaults containing nutrient medium. At the end of each week, two
0.25 m^2 floating PVC frames per vault were placed over the water
surface and the plants falling into these areas were harvested,
fresh weight recorded, and the plants were then returned to the
respective vaults. This procedure was repeated once a week until
the maximum plant density was achieved in the vault and no additional
growth was recorded. At that time, plants were harvested to the
initial density and a sub-sample was analyzed for tissue N and P.
Water samples were obtained at the beginning and at the end of each
week and analyzed for ortho-P.

Experiment II:

This study was conducted to evaluate the optimum nutrient medium
required to achieve maximum P removal by Azolla. Five levels of
nutrient medium (Table 1) containing no N, with each level replicated
three times, was used to culture Azolla. In addition to the nutrient
medium, three types of wastewaters, namely, primary sewage effluent,
secondary sewage effluent, and agricultural drainage effluent were
also evaluated. Primary and secondary sewage effluent were obtained
from the Walt Disney World wastewater treatment facility. Agricultural drainage effluent was obtained from the drainage canals located
in the vegetable farms of Zellwood, Florida. Nutrient composition
of the wastewaters is given in Table 1. Fifty liters of the nutrient
medium or wastewater were placed in 70 l containers. All containers
were placed outdoors. The containers had a surface area of 0.25
m^2. Each of the culture tanks received 400 g m^{-2} (fresh weight)
of plant. Azolla was harvested once a week to their original density
and plant samples obtained each week were analyzed for N and P.
Water samples were obtained at 0, 1, 2, 4, 7, 10, 14, 21, and 28
days after the initiation of the experiment. Water loss due to
evapotranspiration was adjusted to initial volume by adding tap
water. Average solar radiation during the experimental period (August
26 to September 23, 1982) was 3917 k Cals m^{-2} day^{-1}, and average
minimum and maximum ambient air temperatures were 22 and 32 °C,
respectively.

Table 1: Chemical composition of the nutrient medium and wastewater used in the study.

Nutrient	Modified Hoagland nutrient medium					Primary sewage effluent	Secondary sewage effluent	Drainage effluent
	2%	5%	10%	20%	50%			
	——————————————————— mg l^{-1} ———————————————————							
NH_4-N	0.0	0.0	0.0	0.0	0.0	15.5	0.0	0.5
NO_3-N	0.0	0.0	0.0	0.0	0.0	0.0	8.5	0.4
P	1.4	3.4	6.4	12.6		5.6	4.2	1.7
K	4.7	11.8	23.5	47.0	117.5	7.5	6.0	19.5
Ca	4.0	10.0	20.0	40.0	100.0	18.3	14.5	57.0
Mg	1.0	2.5	5.0	10.0	50.0	4.9	4.7	26.0
Fe	0.4	1.0	2.0	4.0	10.0	0.2	0.1	0.1

[+]Fe was added as Fe-EDTA to the nutrient medium.

Minor elements were added at the same concentration to all levels of nutrient medium.

Experiment III:

This part of the study was designed to determine the effect of ambient air temperature on growth and P removal by Azolla cultured in nutrient medium containing no N. Composition of the nutrient medium was similar to that used in Experiment II. Sixty liters of the nutrient medium was placed in 70 l containers having a surface area of 0.25 m^{-2}. Triplicate containers were placed in the growth rooms maintained at a constant temperature of 10, 15, 20, 25 and 30°C. Each of the culture tanks received 200 g m^{-2} (fresh weight) of plant. Nutrient medium was replaced once a week with fresh medium. Azolla plants were weighed once a week until the maximum plant density was recorded. At that time, plants were harvested and a sub-sample was analyzed for N and P. Water samples were taken at the end of each week and analyzed for P.

Experiment IV:

This study was conducted to determine the effect of light on growth and P removal by Azolla cultured in N-free nutrient medium. Composition of the nutrient medium was the same as described in Experiment II. Ten liters of nutrient medium were placed in 12 l containers having a surface area of 500 cm^{-2}. Each of the culture tanks received 200 g m^{-2} (fresh weight) of plants which were allowed to grow for a period of 15 days. All containers were placed in a greenhouse where air temperatures were maintained at approximately the same level as outside the greenhouse. Varying light levels were accomplished by using different layers of shade screens placed over the culture tanks. The treatments include 0, 20, 40, 70, 85, 92, 95, and 99% reduction in light, which represented an average solar radiation of 3156, 2525, 1894, 947, 473, 252, 157, and 32 k Cals m^{-2} day^{-1} , respectively. Water and plant samples were obtained at the end of 15 days and analyzed for P.

Analytical Methods:

Soluble ortho-P in the water samples was determined using an auto-analyzer. Plant samples were dried in a force draft oven at 70°C for a period of 48 hours. Phosphorus in the samples was determined by an autoanalyzer, after digestion with nitric and perchloric acid.

3. Results and Discussion

3.1 Growth Rate. Azolla cultured in nutrient medium (Experiment I) reached maximum density of 105 g dry wt m^{-2} in about 5 weeks during January-February, and during summer months maximum density of about 90 g dry wt m^{-2} was achieved in about 3 weeks. This repre-sents about a 13-fold increase in plant weight in 5 weeks during winter months and about an 11-fold increase in plant weight in 3 weeks during summer months.

Azolla growth rates were significantly influenced by the composition of the nutrient medium or type of wastewater used (Table 2). At the end of the first week and after addition of nutrients, biomass yields were found to be approximately the same for the Azolla cultured in 2, 5, and 10% nutrient medium and yield was significantly decreased when plants were cultured in 20 and 50% nutrient medium. During the second week after the addition of nutrients, growth rates were significantly decreased for Azolla cultured in 2% nutrient medium, primarily due to depletion of nutrients. At higher nutrient concen-tration (5 to 20% nutrient medium), relatively high growth rates (6 to 7 g dry wt m^{-2} day^{-1}) were maintained during the second week after addition of nutrients. At the end of the 4th week, Azolla yields were in the range of 2.5 to 3.9 g dry wt m^{-2} day^{-1} , when cultured in 5 to 50% nutrient medium. Among the wastewaters evalu-ated, Azolla growth during the first week was rapid (8.2 g dry wt m^{-2} day^{-1}) when cultured in primary sewage effluent, followed by secondary sewage effluent (6.5 g dry wt m^{-2} day^{-1}) and drainage effluent (2.7 g dry wt m^{-2} day^{-1}). During subsequent weeks, growth of Azolla was drastically reduced. For all treatments (nutrient medium or wastewaters), Azolla growth was found to be highest during the first week and the yields decreased subsequently as the nutrients were depleted from the water. These results suggest that weekly exchange of nutrients or wastewater in culture tanks can result in maximum biomass yields.

Table 2: Biomass yields of <u>Azolla</u> <u>caroliniana</u> as influenced by the nutrient medium.

Nutrient medium	Weeks after addition of nutrients			
	1	2	3	4
	$\text{------------g dry wt m}^{-2}\text{ day}^{-1}\text{ -------}$			
2% NM	7.9	5.1	2.1	1.0
5	7.4	7.0	5.5	2.5
10	7.0	6.4	6.0	3.8
20	6.8	6.0	4.1	3.8
50	5.6	4.7	4.7	3.9
Drainage effluent	2.7	2.1	1.7	0.0
Secondary sewage effluent	6.5	3.5	0.7	0.4
Primary sewage effluent	8.2	5.8	3.2	1.2
LSD (0.05)	0.7	0.9	1.0	0.9

NM = nutrient medium

Growth rates of <u>Azolla</u> were also found to be influenced by plant density (Figure 1). Optimum plant density to achieve maximum biomass yield was found to be in the range of 30 to 50 g dry wt m^{-2} (0.75 to 1.25 kg fresh wt m^{-2}). At low nutrient concentration, optimum plant densities required to achieve maximum growth will be lower than the values shown above. Other researchers have recommended a wide range of plant densities to achieve maximum yields. For example, Tran and Dao recommended an <u>A</u>. <u>pinnata</u> plant density of 0.5 kg fresh wt m^{-2} which increases to a density of 1 to 1.6 kg fresh wt m^{-2} in winter and 1 to 1.4 kg fresh wt m^{-2} in summer. Talley et al. also used a plant density of 0.5 kg fresh wt m^{-2} in their experiments with <u>A</u>. <u>mexicana</u> and <u>A</u>. <u>filiculoides</u>. Growth rates of <u>Azolla</u> were also influenced by solar radiation and ambient air temperature. Growth of <u>Azolla</u> was not affected when the incoming solar radiation was decreased by up to 60%, but subsequent reduction in solar radiation levels decreased the biomass yields significantly (Table 3). Optimum solar radiation levels to achieve maximum biomass yield were found to be in the range of 1800 to 3200 k Cals m^{-2} day^{-1}. Optimum temperature to achieve maximum growth rate of <u>Azolla</u> was found to be in the range of 20 to 30°C, and growth rates were significantly lower at 10 and 15°C (Figure 2). Lumpkin and Plucknett, summarizing the data published in the Far East also reported an optimum temperature range of 20 to 30°C for maximum growth of <u>Azolla</u> <u>pinnata</u>[2]. Outside this range, growth decreased until the plant began to die at temperature below 5 C and above 45 °C.

Table 3: Effect of light intensity on phosphorus removal by Azolla caroliniana cultured in N-free medium for a period of 15 days.

Lights[‡]	Biomass yield	Phosphorus remaining in the water	Recovery in the plant tissue	
			Nitrogen	Phosphorus
k Cals m^{-2} day^{-1}	-g m^{-2} -	--- mg l^{-1} --	mg N m^{-2}	-mg P m^{-2} -
3,156	44.2	0.08 (97.4)[†]	1,892	123.8
2,525	38.6	0.08 (97.4)	1,641	127.4
1,894	33.0	0.07 (97.7)	1,538	168.3
947	14.9	0.17 (94.5)	592	99.8
473	5.9	0.50 (83.9)	222	57.2
252	3.4	0.88 (71.6)	128	--
157	2.6	1.04 (66.5)	89	--
32	0.4	1.28 (58.7)	12	--

[†]Values in the parenthesis are the percent reduction in P concentration of the water.

[‡]Average solar radiation outside the greenhouse during the study period was 3,945 k Cals m^{-2} day^{-1}.

3.2 Phosphorus Removal. Phosphorus disappearance in the N-free nutrient medium was rapid at all levels of P addition. After the first week's growth period, P concentrations in the water medium decreased from 1.4, 3.4, 6.4, 12.6, and 22.4 mg l^{-} to 0.02, 0.09, 0.64, 1.74, and 3.31 mg P l^{-}, respectively (Figure 3). This represents a P removal rate of 38, 95, 164, 311, and 591 mg P m^{-2} day^{-} for the first week's growth period. At the end of the 4-week growth period, P concentrations decreased to 0.01, 0.02, 0.03, 0.17, and 0.58 mg P l^{-}, respectively, for the nutrient medium with the initial P concentration of 1.4, 3.4, 6.4, 12.6, and 22.4 mg l^{-1}. In the wastewaters evaluated, P concentrations decreased by 62, 90, and 99% during the first week's growth period in drainage effluent, primary sewage effluent, and secondary sewage effluent, respectively (Figure 4). This represents a P removal rate of about 30, 119, and 145 mg P m^{-2} day^{-}, respectively, from drainage effluent, secondary sewage effluent, and primary sewage effluent.

Phosphorus removal from water was influenced by Azolla density in culture tanks (Figure 5). Maximum P removal from the water was observed at relatively low plant densities ($\approx$ 20 g dry wt m^{-2}), and as the plant density in the culture medium was increased, P disappearance from water decreased. At higher densities, <u>Azolla</u> plants tend to overlap other plants forming two to three layers. Increased self-shading of the plants reduces the photosynthetic tissue per unit area to sunlight, thus increasing the detritus plant tissue. The detritus plant tissue upon decomposition releases P into the water, thus reducing the overall P removal efficiency of the system. This can be avoided by increasing the harvesting frequency and maintaining the system at an optimum plant density.

Data in Table 3 shows the effect of solar radiation on P removal from water. Phosphorus removal was approximately the same between treatment receiving solar radiation level of 947 to 3,156 k Cals m^{-2} day^{-1}. Decreasing the light intensity (< 947 k Cals m^{-2} day^{-1}) reduced the P removal efficiency of the system.

3.3 Phosphorus Uptake by <u>Azolla</u>. Phosphorus concentration in <u>Azolla</u> tissue increased with increased level of P in the culture medium. At the end of the first week's growth period (plants harvested after one week) P content of plant tissue was 2.1, 4.9, 6.2, 7.3, and 16.2 mg P g^{-1} of plant tissue, for the culture medium containing an initial concentration of 1.4, 3.4, 6.4, 12.6, and 22.4 mg P l^{-1}, respectively. Depletion of P in the culture medium during subsequent weeks decreased the P content of the plant tissue. At the end of the 4th harvest (4th week after the addition of nutrients) P content of the plant tissues was 0.8, 1.3, 1.6, 3.0, and 4.3 mg P g^{-1} of plant tissue, respectively, for the culture medium with an initial P concentration of 1.4, 3.4, 6.4, 12.6 and 22.4 mg P l^{-1}.

Phosphorus removal due to plant uptake was found to be maximum during the first week's growth period, and decreased during subsequent weeks (Table 4). Plant uptake of P was found to be directly proportional to the P concentration in the culture medium. At low initial P concentrations (1.4 to 6.4 mg P l^{-1}), P uptake by <u>Azolla</u> accounted for about 72 to 95% of the total P removal from the water medium (Table 5). However, plant uptake accounted for about 49 and 44% of the total P removal from the culture medium containing 12.6 and 22.4 mg P l^{-1}, respectively. Low P recovery for the culture medium containing high P concentration indicates the possibility of chemical precipitation. In this study, no attempt was made to account for P tied up in the detritus <u>Azolla</u> (fragmented roots and dead leaves).

Table 4: Phosphorus removal by <u>Azolla</u> <u>caroliniana</u> as influenced by the nutrient medium.

Nutrient medium	Weeks after addition of nutrients			
	1	2	3	4
	---------- mg P m^{-2} day^{-1} ----------			
2% NM	16.7	9.4	1.4	0.9
5	36.7	19.4	5.6	3.3
10	43.5	28.7	21.8	6.1
20	49.4	44.1	22.9	11.3
50	92.1	46.0	43.3	16.3
Drainage effluent	15.2	8.5	4.3	--
Secondary sewage effluent	16.0	3.9	0.6	0.2
Primary sewage effluent	66.8	13.4	5.0	1.1

NM = nutrient medium

Table 5: Mass balance of phosphorus at the 4-week growth period of Azolla cultured in N-free nutrient medium and wastewaters.

Water medium	Phosphorus added	Phosphorus removal		
		Total	Plant uptake	Unaccounted for
	$-mg\ P\ m^{-2}\ -$	---------- % of added P -----------		
Nutrient medium				
2%	270	99.3	94.8	4.5
5%	684	99.4	85.2	14.2
10%	1,276	99.5	72.4	27.1
20%	2,522	98.7	48.5	50.2
50%	4,484	97.5	43.2	54.3
Drainage effluent	334	96.4	106.9	--
Secondary sewage effluent	840	99.8	22.6	77.2
Primary sewage effluent	1,126	99.6	69.3	30.3

3.4 Nitrogen Recovery in the Plant Tissue. Nitrogen content of the plants cultured in N-free nutrient medium was found to be in the range of 3.5 to 3.8% N at the end of the first harvest (one week after the addition of nutrients), and the N content of the plants decreased as the nutrient depletion occurred in the water. At the end of the fourth harvest (four weeks after the addition of nutrients), tissue N content was found to be in the range of 2.36 to 2.92%. At any given harvest period, N content of the plant tissue was found to be approximately the same at all levels of nutrient additions.

Nitrogen concentration of the plant tissue was significantly influenced by the type of wastewater. Nitrogen content of the plants cultured in drainage effluent was found to be approximately the same at all harvests (2.55 to 2.74% N). Low tissue N concentration indicates that N_2 fixation by Azolla-Anabaena relationship was affected by the nutrient composition of the drainage effluents. Nitrogen content of Azolla cultured in sewage effluent was found to be highest during the first harvest, and decreased during subsequent harvest, as the nutrient depletion occurred in the water.

Data in Table 6 show the N recovery rate by Azolla cultured in nutrient medium or wastewater. Nitrogen recovery by Azolla followed similar trends as biomass yields. At the end of one week after addition of N-free nutrients, N recovery in the plant tissue was found to be in the range of 201 to 275 mg N m^{-2} day^{-1} (2.01 to 2.75 kg N ha^{-} day^{-}). During subsequent harvests, N recovery in the plants decreased significantly. At the end of 4 weeks after the addition of nutrients, N recovery in the plant tissue was found to be, in the range of 22 to 111 mg N m^{-2} day^{-} (0.22 to 1.11 kg N ha^{-} day^{-}). Similar trends were also observed for <u>Azolla</u> cultured in drainage effluent (0.45 to 0.75 kg N ha^{-} day^{-}) and sewage effluent (0.06 to 2.92 kg N ha^{-} day^{-}).

Table 6: Nitrogen recovery in the plant tissue as influenced by the nutrient medium.

Nutrient medium	Weeks after addition of nutrients			
	1	2	3	4
	---------- mg N m^{-2} day^{-1} -----------			
2% NM	275	155	61	22
5	268	243	189	73
10	263	205	197	111
20	247	173	119	101
50	201	134	124	102
Drainage effluent	75	57	45	--
Secondary sewage effluent	249	103	14	6
Primary sewage effluent	392	234	100	32
LSD (0.05)	36	30	38	26

NM = nutrient medium

Nitrogen recovery by <u>Azolla</u> was also affected by solar radiation (Table 3). No significant differences in N recovery were observed when solar radiation levels were in the range of 1890 to 3156 k Cals m^{-2} day^{-1}, but further decrese in solar radiation levels decreased the N recovery in the plant tissue.

The major portion of N recovery by plants cultured in N-free medium was probably due to N$_2$ fixation by the <u>Azolla-Anabaena</u> symbiotic relationship, although a small fraction of the N was also derived during decomposition of the detritus plant material and during excretion of ammonia from the plant tissue[5,6].

In conclusion, this study has shown that relatively high growth rates can be maintained when <u>Azolla</u> is cultured in low strength nutrient medium at a residence time of less than 7 days. Growth and P removal rates were found to be influenced by plant density, temperature, and solar radiation. Phosphorus removal by <u>Azolla</u> system was found to be as high as 5.9 kg P ha^{-1} day^{-1}. Phosphorus

removal in this system was found to be due to plant uptake and probably also due to chemical precipitation. Maximum P removal due
to plant uptake was found to be 0.92 kg P ha^{-1} day^{-1}.

Acknowledgements. This paper reports results from a project that
contributes to a cooperative program between the Institute of Food
and Agricultural Sciences (IFAS) of the University of Florida and
the Gas Research Institute (GRI), entitled, "Methane from Biomass
and Waste".

References

1 Cornwell D A, Zoltek J Jr., Patrinely C D, des Furman T and Kim
J I 1977 Nutrient removal by water-hyacinth. J. Water Pollut.
Cont. Fed. 49: 57-65.
2 Lumpkin T A and Plucknett D L 1980 Azolla: Botany, physiology,
and use as a green manure. Econ. Bot. 34: 111-153.
3 Reddy K R, Campbell K L, Graetz D A and Portier K M 1982 Use
of biological filters for agricultural drainage water treatment.
J. Environ. Qual. 11: 591-595.
4 Tran Q T and Dao T T 1973 Azolla: A green compost. Vietnamese
Studies 38, Agric. Problems, Agron. Data 4: 119-127 (Cited by
Lumpkin T A and Pluckentt D L 1980 Econ. Bot. 34: 111-153).
5 Talley S N, Talley B J and Rains D W 1977 Nitrogen fixation by
Azolla in rice fields. In: Alexander Hollaender (ed.) Genetic
Engineering for Nitrogen Fixation. Plenum Press, N.Y. pp. 259-281.
6 Watanabe I, Espinas C R, Berja N C and Alimagno V B 1977 Utilization of the Azolla-Anabaena complex as a nitrogen fertilizer for
rice. Int. Rice Res. Inst. Res. Paper Ser. No. 11.

162

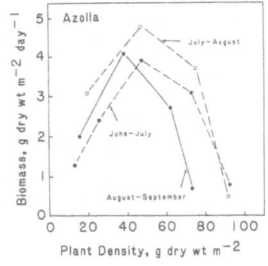 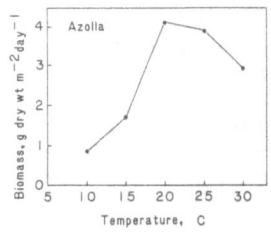

Figure 1. Effect of plant density on growth rate of <u>Azolla</u> <u>caroliniana</u> cultured in N-free nutrient medium.

Figure 2. Effect of ambient air temperature on growth rate of <u>Azolla</u> <u>caroliniana</u> cultured in N-free nutrient medium.

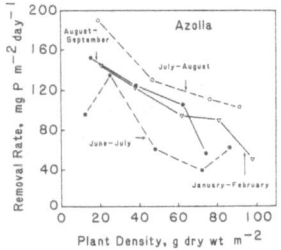

Figure 3. Phosphorus removal from N-free nutrient medium containing <u>Azolla</u> <u>caroliniana</u>.

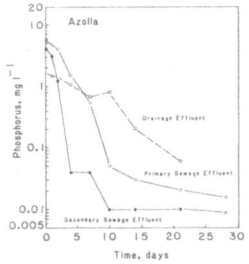 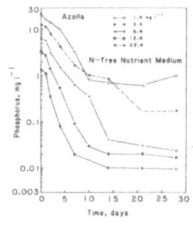

Figure 4. Phosphorus removal from wastewater containing <u>Azolla</u> <u>caroliniana</u>.

Figure 5. Effect of plant density on P removal by <u>Azolla</u> <u>caroliniana</u> system.

15. Utilization of radioactive phosphorus (^{32}P) by <u>Azolla-Anabaena</u> and its transfer to rice plants

María José Amstalden Moraes Sampaio*, Marli de Fátima Fiore** and Alaides Puppin Ruschel*

* Centro de Energia Nuclear na Agricultura - CENA, USP/CNEN, Av. Centenário s/n - C.P. 96, Piracicaba, SP, Brazil
** Centro Nacional de Pesquisas de Arroz e Feijao EMBRAPA, BR 153, km 4, C.P. 179, Goiania, GO, Brazil

Key words   <u>Azolla-Anabaena</u>   Brazil   Fertilization   Radioactive Phosphorus   Rice   Soils

Summary

Phosphorus is a limiting nutrient for rice production in Brazil. Greenhouse studies with four species of <u>Azolla</u> fertilized with ^{32}P showed that the fern helped the release of P from the fertilizer, increasing rice dry weight.

1. Introduction

The N_2-fixing potential and agronomic importance of the <u>Azolla-Anabaena</u> complex have been recognized, particularly in relation to the soil fertility in paddy rice cultivation[2,4,6]
   In Brazil, a large area (6.10^6 ha) is utilized for rice cultivation, 12% of which is under waterlogged conditions and is responsible for 24.6% ($2.1\ 10^6$ ton) of the total production.
   Phosphorus is one of the most limiting nutrients for upland rice production in brazilian soils. The problem is not so serious in paddy areas but it has been shown that P-fertilization increases rice production, since significant responses in yield were obtained when phosphorus was added up to 88 kg P ha^{-1}[3]. The growth of <u>Azolla</u> for incorporation as green manure in paddy rice cultivation usually needs the addition of phosphorus and, as detailed information on release of P from <u>Azolla</u> is not yet available[5], the main objective of this paper was to follow the transfer of P from ^{32}P-<u>Azolla</u> to rice plants after its incorporation in the soil.

2. Materials and Methods

2.1 <u>Azolla</u>.   Four species of the aquatic fern <u>Azolla</u> were used. <u>A. caroliniana</u>, <u>A. filiculoides</u>, <u>A. microphylla</u> and <u>A. pinnata</u> were grown under greenhouse conditions on diluted (1:3) Hoagland's solution. ^{32}P-labelled <u>Azolla</u> was obtained by growing ferns on the same medium with the addition of 62.5 uCi.l^{-1} of NaH^{32}PO$_4$ (carrier-free) obtained from IPEN/USP-SP.

2.2 Soil.   Alluvial soil was collected from the experimental paddy area at ESALQ campus.   Its main characteristics were: pH 5.4; C% 2.17; total P:55.4 ug.g^{-1}, N% 0.085.

2.3 Incubation conditions. Soil samples (500 g) were put into plastic pots after each sample had been thoroughly mixed with Azolla quantities (2.45 g fresh weight pot^{-1}, sufficient to give 40 Kg N ha^{-1}) or had received ^{32}P-fertilizer (6.2 uCi pot^{-1} in 20 mg P as Ca (H$_2$PO$_4$)$_2$ .H$_2$O). A series without the addition of Azolla or phosphorus served as the control. Pre-germinated rice seeds (IAC-435) were planted and a week later pots were flooded with tap water and kept like that until the harvesting of the experiment. The pots were covered with aluminum foil to prevent algal growth. Each treatment had four replicates.

2.4 Chemical determinations. Samples were dried out at 60°C, digested and prepared to be analyzed in a Plasma Analyzer.

3. Results and Discussion

Rice leaves showed an increase in dry matter (Table 1) when Azolla was incorporated in the soil as compared to the treatment where only P-fertilizer was added. However, this increase did not correlate with an increase in their P content (the increase was probably due to the N content of the Azolla incorporated or other factors not discussed). Root dry weight (Table 2) was higher in treatments with Azolla incorporation than in that with P-fertilizer addition only, the same occurring with total P, which is in accordance with the findings of Banzatto and Azzini[1] that P stimulates the growth of rice roots[1].

An effect of Azolla fertilization and availability of phosphorus, released to the soil during decomposition , and used by the plants is seen when %P derived from the fertilizer (%Pdff) was analyzed. Plants that received only Azolla showed values between 36.6 and 65.5%P in leaves (Table 1) and 24.4 and 45.5%P in roots (Table 2), while only 8.1 and 7.2%P respectively came from P-fertilizer when this was the only source applied to the soil. %Pdff in the treatment where both Azolla and inorganic P were added demonstrated that Azolla incorporation helped to release P from the fertilizer as compared to addition of inorganic phosphorus only. This increase of available P when Azolla was incorporated in soil might be related to a solubilizing effect of organic acids produced as a result of its decomposition[5]. It is therefore clear that plants did not show a phosphorus deficiency during the growth period of 48 days that could affect growth when only Azolla was used as the P-source.

% P-utilization (Table 3) was higher in rice (leaves and roots) from treatments that received Azolla incorporation than in plants from (Azolla + P-fertilizer) or P-fertilizer treatments. By comparing the two last treatments, it is seen that %P utilization was higher in the presence of Azolla than in its absence.

Differences were observed among the various species of Azolla studied. Relative efficiency of P-labelling (Table 4) was higher in A. pinnata than others. %Pdff (leaves and roots) in the treatment with Azolla incorporation was higher for A. caroliniana and A. microphylla than for A. filiculoides and A. pinnata (Tables 1 and 2).

As shown by Saha et al[5] other parameters like electro-chemical and chemical properties of the soils, can change when Azolla is incorporated to the system and therefore a further study of availability of nutrients, mainly at later stages of growth (when tillering for instance, more phosphorus is required) needs to be carried out.

References

1 Banzatto N V and Azzini L E 1971 Guia da Producao Rural, Coopercotia 1971. In: Nutricao Mineral e Adubacao de Plantas Cultivadas. Pioneira (ed.) SP, Brazil. 33-34.
2 Dao T T and Tran A T 1979 Use of Azolla in rice production in Vietnam. In: Nitrogen and Rice. International Rice Research Institute, Los Banos, Philippines. 395-405.
3 Fageria N K 1980 Influencia da aplicacao de fósforo no cresci-mento, producao e absorcao de nutrientes do arroz irrigado. Rev. Bras. Ci. Solo 4, 26-31.
4 Rains D W and Talley S N 1979 Uses of Azolla in North America. In: Nitrogen and Rice. International Rice Research Institute, Los Banos, Philippines. 419-430.
5 Saha K C, Panigrahi B C and Singh P K 1982 Effect of blue-green algae or Azolla additions on the nitrogen and phosphorus avail-ability and redox potential of a flooded rice soil. Soil Biol. Biochem. 14, 23-26.
6 Singh P K 1979 Use of Azolla in rice production in India. In: Nitrogen and Rice. International Rice Research Institute, Los Banos, Philippines. 407-418.

Table 1: Effect of _Azolla_ and/or P-fertilizer addition on dry weight, total P and %Pdff on rice leaves.

Treatments	Dry weight (mg pot^{-1})	Total P (mg)	%Pdff
^{32}P– _A. caroliniana_	588	2.7	61.8
^{32}P– _A. filiculoides_	608	2.9	46.1
^{32}P– _A. microphylla_	658	2.4	65.5
^{32}P– _A. pinnata_	672	3.2	36.6
^{31}P– _A. caroliniana_ + ^{32}P–fertilizer	522	2.5	14.9
^{31}P– _A. filiculoides_ + ^{32}P–fertilizer	560	2.3	17.1
^{31}P– _A. microphylla_ + ^{32}P–fertilizer	663	2.9	14.5
^{31}P– _A. pinnata_ + ^{32}P fertilizer	717	2.5	18.6
^{32}P– Fertilizer only	445	2.2	8.1

Table 2: Effect of _Azolla_ and/or P-fertilizer addition on dry weight, total P and %Pdff on rice roots.

Treatments	Dry weight (mg pot^{-1})	Total P (mg)	%Pdff
^{32}P– _A. caroliniana_	464	0.74	45.0
^{32}P– _A. filiculoides_	506	0.45	36.8
^{32}P– _A. microphylla_	599	0.96	45.5
^{32}P– _A. pinnata_	576	0.86	24.4
^{31}P– _A. caroliniana_ + ^{32}P–fertilizer	418	0.65	6.9
^{31}P– _A. filiculoides_ + ^{32}P–fertilizer	642	0.96	6.2
^{31}P– _A. microphylla_ + ^{32}P–fertilizer	542	0.76	5.4
^{31}P– _A. pinnata_ + ^{32}P–fertilizer	582	0.81	5.8
^{32}P– Fertilizer only	242	0.39	7.2

Table 3: Percentage utilization of fertilizer by rice.

Azolla	Without inorganic ^{32}P-fertilizer + ^{32}P-Azolla		With inorganic ^{32}P-fertilizer + ^{31}P-Azolla	
	leaves	roots	leaves	roots
A. caroliniana	8.8	1.7	1.9	0.1
A. filiculoides	5.5	0.7	2.0	0.2
A. microphylla	5.3	1.5	2.3	0.1
A. pinnata	6.4	1.2	2.3	0.1
None	–	–	0.9	0.1

Table 4: Relative efficiency of ^{32}P labelling of four species of Azolla.

A. caroliniana	0.67
A. filiculoides	0.53
A. microphylla	0.51
A. pinnata	1.00

16. A study on the availability of biologically fixed atmospheric dinitrogen by the Azolla-Anabaena complex to flooded rice crops.

M. H. Mian
Assistant Professor
Department of Soil Science
Bangladesh Agricultural University
Mymensingh, Bangladesh

and W. D. P. Stewart
Professor and Head
Department of Biological Sciences
The University of Dundee
DD4 4HN, Scotland, U.K.

Key words Azolla   decomposition   gaseous loss ^{15}N tracer   N release   rice

Summary

A ^{15}N tracer study was planned to obtain information on the following aspects: (a) the rate of decomposition of Azolla applied to a flooded rice soil, (b) uptake of mineralized Azolla-N by the standing rice crop and its effect on rice growth and (c) fates of remaining mineralized part of Azolla-N in the flooded rice soils.

^{15}N tracer shows that Azolla applied to flooded soil without rice plants started to decompose within a few days of its addition into the flooded soil and NH_4^+-N continued to accumulate in soils for about a month. Afterwards NH_4^+-N gradually disappeared during the next month but did not reappear as NO_3^--N. However, gas analysis revealed that the NH_4^+-N was lost as N_2 through nitrification-denitrification processes. Hardly 2% of the mineralized Azolla-N remained as NH_4^+-N and NO_3^--N in the soil after 60 days.

In another set of experiments with rice plants it is found that about 36% of the total Azolla-N added at the start was released in 60 days and of that about 71% was assimilated by the rice plants, 2% remained in soil as NH_4^+-N and NO_3^--N, and 27% was lost to the atmosphere as gas.

From this study we conclude that: (i) the flooded rice plants received a substantial amount of nitrogen from decomposing Azolla and the rice dry matter yield was significantly increased; (ii) only traces of mineralized Azolla-N remained in soil; (iii) about 27% was lost as gas; (iv) in absence of a rice crop the gaseous loss of the mineralized Azolla-N was almost total; and (v) such loss could be minimized if planted with a rice crop within 2-3 weeks after application of Azolla to the flooded soil.

1. Introduction

The continuous increase in human population has put tremendous pressure on increasing food production. About half of the world's population live on rice and over 90% of the rice is grown in

south-east Asia. Colombo et al.[4] reported that Asia must double
its rice production from its 1974 level by 1994 in order to feed
its increasing population. This achievement will depend on the
amount of nitrogen available since 1.7 kg N is required to produce
100 kg of brown rice[8], a situation unlikely to be realized with
the use of chemical fertilizers due to its high cost and general
unavailability.. Consequently attention has been diverted toward
the possible use of biofertilizers. The Azolla-Anabaena symbiosis
represents potentially ideal biofertilizer for rice production due
to its high nitrogen-fixing ability and rapid growth. Many reports
are available on the positive effects of Azolla application in rice
culture[5,6,7,11,12]. Most of these demonstrated a higher rice yield
when rice was grown in the presence of Azolla. However, the availability of Azolla-N to the rice crop and the fate of mineralized
Azolla-N in flooded rice soils is not adequately documented. This
study was undertaken in order to obtain a better understanding of
these processes.

2. Materials and Methods

2.1 Soil. A Bangladesh loam soil from 0-15 cm depth (pH 7.2, 0.05%
total N and 0.97% organic matter) was used in the study.

2.2 Nitrogen biofertilizer. Azolla caroliniana Willd. collected
from the Dundee University Botanic Garden was labelled with [15]N
by growing them in the Na[15]NO_3^- enriched medium of Peters and
Mayne[9] for three weeks. The fronds were then thoroughly washed
and grown for another two days in an N-free medium in order to exclude
any traces of adhered but not assimilated Na[15]NO_3. The harvested
and air-dried fronds were used as the source of nitrogen.

2.3 Experimental procedure

   (a) Without rice plants

   The air-dried Azolla biomass supplying 200 ug N g^{-1} soil was
mixed thoroughly with 80 g soil in one liter capacity glass bottles.
A 4 cm flood water layer with an air space above was made in each
bottle. All bottles were sealed with suba seals and incubated in
a growth cabinet at 25°C (12 h, light) and 18°C (12 h, dark) temperatures. NH_4^+-N and NO_3^--N in the incubated soils were determined according to Bremner[1]. Gas samples drawn from the bottles before
each determination were analyzed in a VG MM 601 micromass spectrometer
for estimating gaseous loss of nitrogen.

   (b) With rice plants

   Eighty milligrams of Azolla-N was added in each pot which contained
1 kg of soil. Each treatment was duplicated. Four 16 days old
rice seedlings (var IR8) were transplanted to each pot. A 4 cm
flood water was maintained in each pot throughout the experimental
period by the addition of distilled water. The pots were kept in
a growth cabinet at 25°C (12 h, light) and 18°C (12 h, dark) cycles.
The plants were harvested at 60 days of growth, oven-dried at 65°C
and constant weights of the shoots and roots were recorded.

Nitrogen in shoots and roots, and total nitrogen in the soils were determined according to Bremner[2]. NH_4^+-N and NO_3^--N in the soils were analyzed as in (a) above.

2.4 ^{15}N determinations. ^{15}N present in the samples following the determinations of NH_4^+-N, NO_3^--N and total nitrogen were estimated according to the method described by Bremner[3].

3. Results

Table 1 shows that nitrogen released from the decomposing Azolla fronds accumulated in the flooded soils as NH_4^+-N up to 30 days of incubation. After 30 days the amount gradually disappeared and at 60 days practically no NH_4^+-N was present in the soils. In Azolla-treated soils at 60 days NO_3^--N was negligible compared to the amount of NH_4^+-N which had disappeared. In control soils NH_4^+-N did not accumulate and NO_3^--N continuously accumulated in the soils up to 60 days. Since in the Azolla-treated soils the situation was completely different, ^{15}N tracer was used to monitor the fates of the mineralized Azolla-N.

The measured quantities of the available fractions of Azolla-N by ^{15}N tracer are shown in Table 2. Here it is seen that about 20.8 mg of mineralized Azolla-N was present in soils as NH_4^+-N at 30 days of incubation whereas at 60 days only 2.0 mg remained. In both the soils NO_3^--N was negligible. The gaseous loss of the mineralized Azolla-N was only 7.2 mg at 30 days (about 3.6% of the total amount mineralized) whereas it was 62.0 mg at 60 days (about 31.0% of the total amount mineralized). Therefore, it appears that including gaseous loss the total amount of mineralized Azolla-N amounted to about 14% and 33% of the total amount added at the beginning.

The availability of the mineralized Azolla-N to the flooded rice plants was also tested. Table 3 presents data on the dry matter yield of the harvested rice plants at the end of 60 days of growth only although the rice plants of 7, 15 and 30 days of growth were also tested for ^{15}N incorporation. ^{15}N was detected even in 7 days old transplanted rice plants grown in Azolla treated soil. The shoot dry matter yield was found to be 539 mg pot^{-1} in case of control whereas it was 883 mg pot^{-1} and 1093 mg pot^{-1}, respectively, for the additions of 36 ug Azolla-N g^{-1} soil and 80 ug Azolla-N g^{-1} soil, respectively. Similarly root yield also increased. Table 3 also shows that the increase in yield was directly dependent upon the increased amount of Azolla-N assimilated by the rice plants. Where the rice plants of the control pot received only 13 mg N, the rice plants grown in the soils treated with Azolla-N at the rate of 36 and 80 ug g^{-1} soil received 21.9 mg N and 31.8 mg N, respectively. That is, the rice plants grown in the Azolla-treated pots received about 68% and 145% more nitrogen, respectively, than the plants grown in the control pots.

The use of ^{15}N tracer enabled us to differentiate the amount of nitrogen received by the rice plants from Azolla and soil (Table 4). The control plants received all their nitrogen (15.1 mg) from soil. The soil derived nitrogen fraction in the rice plants grown in the Azolla-treated soil was almost same as that was found in the rice plants grown in the control soil. The extra amount of

nitrogen was obtained by the rice plants from the added <u>Azolla</u> fronds.
Table 5 shows a complete picture of the states of mineralized
<u>Azolla</u>-N in the flooded rice soils. In 60 days 16.0 mg of <u>Azolla</u>-N
was released out of 36 mg N added at the start. Of the mineralized
amount, 73% was assimilated by the growing rice plants, 1% remained
in soil as NH_4^+ -N plus NO_3^- -N and 26% was lost as gas. In the case
where 80 mg <u>Azolla</u>-N was added, 28.6 mg N was released of which
71% was found assimilated in the rice plants, 2% remained as
NH_4^+-N and NO_3^--N in the soil and 27% was lost as gas.

4. Discussion

The results show that decomposition of <u>Azolla</u> biomass started within
few days of its addition into the flooded soil with the release of
NH_4^+ -N. The released NH_4^+ -N continued to accumulate in soil during
the first one month but gradually disappeared in the next month
instead of being maintained by a continuous decomposition of <u>Azolla</u>
biomass. The NH_4^+-N which disappeared was not found in the soil
NO_3^- -N pool. However, gas analysis revealed that after 30 days,
oxidation of the generated NH_4^+-N occurred and the NO_3^--N thus produced
could not stay in soil due to gradual reduction into gaseous nitrogen
(Table 2). The first determination at 30 days detected 20.8 ug
NH_4^+ -N, 0.4 ug NO_3^- -N and a gaseous loss of 7.2 ug N_2 g^- soil in
a place where the second determination detected only 2.0 ug NH_4^+-N,
1.0 ug NO_3^- -N and a gaseous loss of 62 ug N_2 g^- soil. The sum total
of mineralized <u>Azolla</u>-N at 30 days amounted to 28.4 ug g^- soil
but it was 65.0 ug g^- soil at 60 days. Thus it is certain that
at about 30 days of the experiment nitrification–denitrification
reactions became sufficiently significant to diminish the mineralization of NH_4^+-N. Similar losses of added nitrogen from flooded soil
through nitrification–denitrification processes has been reported
by Reddy and Patrick [6] .
The situation was different when the experiment was repeated in
presence of a rice crop. The gaseous loss of the mineralized
NH_4^+ -N was only 27% whereas such loss was over 96% when there were
no rice plants present. Of the total amount of <u>Azolla</u>-N released
as NH_4^+ -N, 71% was assimilated by the rice plants and gaseous loss
was reduced. The amount of released nitrogen from decomposing <u>Azolla</u>
fronds was sufficient to significantly increase rice yield.

References

1 Bremner J M 1965a Inorganic forms of nitrogen. <u>In</u>: Methods
  of soil analysis, part 2 (eds. C. A. Black et al.). Am. Soc.
  Agron. pp. 1179–1237. Madison, U.S.A.
2 Bremner J M 1965b Total nitrogen. <u>In</u>: Methods of soil analysis
  part 2 (eds. C. A. Black et al.). Am. Soc. Agron. pp. 1149–1178.
  Madison, U.S.A.
3 Bremner J M 1965c Isotope ratio analysis of nitrogen in N–15
  tracer investigations. <u>In</u>: Methods of soil analysis, part 2
  (eds. C. A. Black et al.). Am. Soc. Agron. pp. 1256–1286. Madison,
  U.S.A.
4 Colombo U, D G Johnson and T Shields 1977 Expanding food production in developing countries: rice production in south and
  south–east Asia. Report of the Trilateral Food Task Force

presented at the Trilateral Commission Meeting, Bonn, FRG.

5 Dao T T and Q T Tran 1979 Use of *Azolla* in rice production in Vietnam. In: Nitrogen and rice. IRRI, Los Banos, Laguna, Philippines. pp. 395–405.

6 FAO 1978 China: *Azolla* propagation and small-scale biogas technology. FAO Soils Bull. 40: 29–40.

7 Lumpkin T A and D L Plucknett 1980 *Azolla*: Botany, physiology and use as a green manure. Economic Botany 34: 111–153.

8 Park J K, S T Lee and S H Choi 1969 Studies on the growth condition and nutriophysiology on paddy rice in high and low productive paddy field. Res. Rep. Off. Rural Deve. (Suwon) 12(3): 75–95.

9 Peters G A and B C Mayne 1974 The *Azolla-Anabaena* azollae relationship. I. Initial characterization of the association. Plant Physiol. 53: 813–819.

10 Reddy K R and W H Patrick Jr 1980 Losses of applied ammonium ^{15}N, Urea ^{15}N, and organic ^{15}N in flooded soils. Soil Sci. 130: 326–330.

11 Singh P K 1979 Use of *Azolla* in rice production in India. In: Nitrogen and rice. IRRI, Los Banos, Laguna, Philippines. pp. 407–408.

12 Talley S N and D W Rains 1980 *Azolla filiculoides* Lam. as a fallow-season green manure for rice in a temperate climate. Agron. J. 72: 11–18.

Table 1: Decomposition of ^{15}N-labelled _Azolla_ fronds with the release of NH_4^+-N and NO_3^--N in flooded rice soil in 60 days.

Treatments	ug.g^{-1} soil			
	0	15	30	60
Check				
NH_4^+-N	4.2	3.1	1.0	0
NH_3^--N	7.6	11.0	14.1	16.5
Azolla				
NH_4^+-N	4.2	22.3	20.5	0
NH_3^--N	7.6	2.0	5.3	2.4

Table 2: Estimation of different forms of mineralized _Azolla_-N in flooded rice soils by ^{15}N tracer technique.

Time of incubation (days)	Amount of Azolla-N added at the start (ug^{-1} soil)	Amount of Azolla-N mineralized ug.g^{-1} soil)			
		NH_4^+-N	NO_3^--N	N_2-N gas	Total
30	200	20.8 (10.4)	0.4 (0.2)	7.2 (3.6)	28.4 (14.2)
60	200	2.0 (1.0)	1.0 (0.5)	62.0 (31.0)	65.0 (32.5)

The figures within the brackets represent percentages for the corresponding values.

Table 3: Effects of _Azolla_ application on the growth of flooded rice plants grown for 60 days in pots.

Treatments	Dry matter yield ($mg\ pot^{-1}$)	% increase over control	Total N yield ($mg\ pot^{-1}$)	% increase over control
Check				
Shoot	539	0	13.0	0
Root	128	0	2.1	0
$36\ \mu g$ Azolla-N g^{-1} Soil				
Shoot	883	64	21.9	68
Root	187	46	3.7	76
$80\ \mu g$ Azolla-N g^{-1} Soil				
Shoot	1093	103	31.8	145
Root	241	88	4.1	95

Table 4: Assimilation of nitrogen released from _Azolla_ by the flooded rice plants in 60 days.

Azolla-N added ($mg\ pot^{-1}$)	Total N uptake by the plants ($mg\ pot^{-1}$)	Amount of N derived from fertilizer ($mg\ pot^{-1}$)	% of total uptake	Amount of N derived from soil ($mg\ pot^{-1}$)	% of total uptake
Check	15.1	---	---	15.1	100
36	21.4	10.3	48	11.1	52
80	35.8	20.4	57	15.4	43

Table 5:   Distribution of mineralized <u>Azolla</u>-N in plants and soil.

Azolla-N added (mg pot^{-1})	Total amount of N mineralized (mg pot^{-1})	Amount of N assimilated by the rice plants (mg pot^{-1})	NH$_4^+$-N and NO$_3^-$-N left in soil (mg pot^{-1})	Amount of N lost (mg pot^{-1})
36	16.0 (100)	11.6 (73)	0.2 (1)	4.2 (26)
80	28.6 (100)	20.4 (71)	0.5 (2)	7.7 (27)

The figures within the brackets represent percentages for the corresponding values.

THIRD SECTION: COUNTRY STUDIES

17. Azolla for rice production in Thailand

Chob Kanareugsa, Prayoon Swatdee, Somjit Khonthasuvon,
Laddawan Loudhapasittiporn and Nantharat Suphakumnerd
Rice Division, Department of Agriculture
Ministry of Agriculture and Cooperatives
Thailand

Key words   Azolla   Fertilizer   Nitrogen   Phosphate   Rice   Thailand

Summary

Thai farmers cannot apply chemical nitrogen fertilizers in the quantities recommended by the government because the price of rice is
very low while the price of commercial fertilizers is relatively
expensive. Azolla is very suitable as a source of nitrogen for
rice in Thailand for the following reasons: the short time necessary
for obtaining a substantial supply of nitrogen (as high as 3 kg
N/ha/day), their N content (4-5%) and their tolerance of acidity
and high temperature.
Azolla will not grow unless phosphate is supplied at the seeding
time. Farmers must seed Azolla about 20 days before transplanting
rice but in rainfed areas, there is insufficient time to seed Azolla
before transplanting, due to the short period of water kept in the
field. However, Azolla may be seeded between hills of rice plants
immediately after transplanting. Because it is uneconomical to
plow Azolla into the soil, Azolla may be allowed to decompose and
release nitrogen slowly by natural decomposition processes.
One crop of Azolla, whether added before or after transplanting
gave yields which were almost equivalent to the 30 kg/ha of chemical
nitrogen fertilizer. Two crops of Azolla seeded before and after
transplanting gave the similar yield as in the 60 kg N/ha fertilized
plot. There was no difference in grain yield, whether Azolla seeded
after transplanting of rice was incorporated or not.

1. Introduction

Rice serves as the main foodstuff of the Thai people and the surplus
is exported and commands the highest value of Thailand's overseas
trade. At present, Thailand improved the new technology and science
for growing rice under farmer's condition. The use of modern rice
varieties, fertilizers, and pest and diseases control is practiced.
From experiments as well as the results of government demonstrations
on farmer's fields, it was shown that the paddy yields of 4 to 6
tons per hectare can also be produced economically in Thailand.
In 1982, Thailand had a population about 48 million people. Yet
there is still a surplus of rice for export, but in the near future
there will be nothing left to export. Therefore research and the
direct application of those results to the farmer's production will
be imperative in the very near future.
Azolla is a very suitable source of nitrogen for rice in Thailand
for the following reasons:

1. The short length of the time necessary for obtaining a substantial supply of nitrogen (as high as 3 kg N/ha/day).

Owing to the unpredictable rainfall pattern in Thailand, the time required for green manure crops should be as short as possible. Azolla meets this requirement best among all the green manure crops grown in Thailand. Seeding of 50 gm of fresh Azolla produces 20 to 40 kg of fresh Azolla containing 4-5% N thereby supplying about 3 kg nitrogen per hectare per day of the rice.

2. A high tolerance of acidity and high temperature

Azolla can grow in an acid medium of pH 3.5. This is a great advantage particularly in acid sulfate soil area, for example in Rangsit. The surface water of rice fields develops an acidity as strong as pH 3.8. On the other hand, growth of Azolla is retarded under high temperature. However, it is fortunate that Azolla pinnata, Bangkok strain, which is found widespread in little ponds in Thailand, can grow all the year round even under a high temperature condition.

3. Suitability to low fertility soil

In the North-Eastern Region, low fertility soils are mostly sandy in texture and low in free iron oxide which cause phosphate precipitation in water. This is very important because phosphate is the only element necessary to be supplied for Azolla growth.

Azolla requires the same essential nutrients as the higher plant and particularly they need phosphorus. Azolla will not grow unless phosphate is supplied at the seeding time. Split phosphorus applications are recommended as the best method for growing Azolla.

2. Materials and Methods

Experiments on the utilization of Azolla for rice production have been conducted in Thailand since 1976. The results of the experiments in 1977 (Table 1) revealed that rice yield tended to increase as the application of phosphate fertilizer to Azolla was split. The number of times of P-Split application did not affect paddy yield. However, when fresh weight of Azolla was considered the best treatment was five time-split applications. This was possibly an advantage for the following crop because the residue of Azolla matter accumulated and improved the soil.

In previous studies Azolla was seeded at about 20 days before transplanting rice, thus the time for rice culture took much longer. In rainfed areas, rice farmers do not have enough time to seed Azolla before transplanting, due to the short period of water in the field. Accordingly in other experiments, Azolla was seeded in the rice field immediately after transplanting . In these experiments the Azolla dies off under the rice canopy, releasing nitrogen into the paddy and plowing is therefore unnecessary. This practice increased the rice yield. In these experiments Azolla was inoculated only once. In other experiments a combination of pre-transplanting and post-transplanting inoculation of Azolla was tried at eight locations in 1979[2]. The treatment descriptions and grain yields are given in Tables 2 and 3 respectively.

3. Results

The results indicate that one crop of Azolla (whether before or after transplanting) followed by incorporation gave yields which were almost equivalent to the 30 kg/ha chemical fertilized plot from the average of each treatment. Two crops of Azolla, seeded before and after transplanting, gave a similar yield as in the 60 kg N/ha fertilized plot. There was no difference in grain yields, whether or not Azolla seeded after transplanting the rice was incorporated.

Under the INPUTS project of the East-West Center and the Rice Division, Thailand, the field experiment was set up at Chumpae Rice Experiment Station. The amount of Azolla to apply was determined by calculation, assuming that 3.0% on the dry weight was N (Table 4).

It has been reported that the fungus Myrothecium sp., common to many different plants and are frequently isolated from soil and textiles, will rapidly kill Azolla. The disease symptom is scattering with distinct white spots of fungus mycelium ·

4. Discussion

4.1 Effects of organic matter. Investigation on effect of rice bran and compost which compared with superphostate on Azolla growth revealed that applications of different types of organic matter increased Azolla growth. Azolla appeared dark green and bigger fronds than chemical fertilizer treated in the third crop with bran was treated in the first crop only (Table 5).

4.2 Effect of type of phosphate applied on Azolla growth and grain yield. Study on comparison among kinds of phosphate fertilizer which combined with nitrogen. The experiments were conducted at Sakon Nakhon and Pimai Rice Experiment Stations. The result of 6-24-24 fertilizer utilization showed the satisfactory effect. Azolla growth and grain yield responded to 6-24-24 (or 1.3-4-4 kg/rai) fertilizer the same as superphosphate (0-4-4) application (Tables 6 and 7).

The results of all previous experiments, although showing promise, are still inconclusive. As the trend of results indicate, the potential of Azolla as a supplementary source of nitrogen requires further researches in future. The finding of Myrothecium sp. as the causative organism of Azolla infection also brought to further interesting study. The fungus may not only be parasitic to Azolla but also to the rice. Ecological study as well as control measure to this fungus should also be considered.

References

1 Swatdee P et al 1978 Utilization of Azolla for rice cultivation. Rice Fertilization Research Branch, Rice Division, Dept. of Agriculture, Thailand (Roneo).
2 Swatdee P et al 1979 Split application of phosphate fertilizer on Azolla and rice yield in acid sulfate soil. Rice Fertilization Research Branch, Rice Division, Dept. of Agriculture, Thailand. (Unpublished Report).

182

3 Disthaporn S et al   1981   Study on "rotten" disease of _Azolla_. Plant   Pathology Division, Dept. of   Agriculture,   Thailand. (Unpublished Report).

Table 1: Rice yield as affected by P – split application to Azolla.

Treatment number	Description	Fertilizer application for Azolla growth (kg/ha) N – P₂O₅ – K₂O	Grain yield (kg/ha)	Weight of fresh Azolla (ton/ha)
1	No Azolla	0–25–25	1406 c	–
2	No Azolla	37.5–25–25	2393 a	–
3	Azolla (P-non split)	0–25–25	1425 c	4.28 d
4	Azolla (P-split 2 times)	0–25–25	2031 ab	13.88 c
5	Azolla (P-split 4 times)	0–25–25	1762 bc	16.11 bc
6	Azolla (P-split 5 times)	0–25–25	2118 ab	21.27 a
7	Azolla (P-split 10 times)	0–25–25	2281 ab	17.76 abc
8	Azolla (P-split 20 times)	0–25–25	2619 ab	19.44 ab

C.V. = 16.5%

Means with similar letters are not significant at 5% level.

Table 2: Description of treatments.

Treatment number	Description of treatment	Fertilizer application	
		Pre-transplanting P_2O_5 (kg/ha)	Post-transplanting $N-P_2O_5-K_2O$ (kg/ha)
1	No Azolla	0	0-0-0
2	No Azolla	0	0-25-25
3	No Azolla	0	30-25-25
4	No Azolla	0	60-25-25
5	Azolla, seeded before transplanting and plowed in	25	0-0-25
6	Azolla, seeded at transplanting and plowed in	0	0-25-25
7	Similar to #6 but not plowed in	0	0-25-25
8	Similar to #5	25	30-0-25
9	Similar to #6	0	30-25-25
10	Azolla, seeded before and at transplanting time and plowed in both	25	0-25-25
11	Azolla, seeded before transplanting and plowed in, and seeded at transplanting time	25	0-25-25
12	Azolla, seeded before transplanting and plowed in, and applied fertilizer to the rest of Azolla	25	0-25-25

Table 3: Effect of Azolla on rice yield (ton/ha) in 1979 season.

Treatment	Location								Mean
	SRN	SKN	CPA	PMI	PAN	UBN	RST	KGT	
1	1.7 e	1.5 d	2.1 e	3.9 def	2.7 d	2.0 e	2.2 h	1.7 c	2.2
2	2.0 de	1.7 bcd	2.4 cd	3.3 f	3.0 d	2.1 e	2.5 gh	1.8 c	2.4
3	2.7 bc	2.0 abcd	2.7 abc	4.5 bcd	3.8 ab	2.8 ab	3.3 cde	2.3 b	3.0
4	3.0 ab	2.5 a	2.9 a	5.0 ab	4.1 a	3.1 a	4.3 a	2.8 a	3.5
5	2.6 bc	1.8 bcd	2.4 cd	4.2 cde	3.3 c	3.0 ab	3.0 ef	2.8 a	2.9
6	2.8 bc	1.7 bcd	2.4 cd	3.7 ef	3.4 c	2.6 cd	2.8 fg	2.3 b	2.7
7	2.3 cd	1.8 bcd	2.5 bcd	3.9 def	3.4 c	2.4 d	3.2 def	2.3 b	2.7
8	3.3 a	2.1 abc	2.7 abc	5.3 a	3.8 ab	2.9 ab	4.1 ab	3.0 a	3.4
9	2.9 ab	2.3 ab	2.8 ab	4.9 ab	4.0 ab	2.8 ab	3.5 cd	2.5 b	3.2
10	3.5 a	2.0 abcd	2.7 abcd	4.5 bcd	3.9 ab	2.9 ab	3.7 bc	3.0 a	3.3
11	2.8 bc	1.8 bcd	2.6 abcd	4.7 abc	3.8 ab	3.0 ab	3.7 bc	2.9 a	3.2
12	3.0 ab	2.0 abcd	2.7 abc	4.6 bcd	3.8 ab	2.9 ab	3.4 cde	2.8 a	3.1
C.V. (%)	12.5	18.4	7.8	10.4	5.6	6.5	7.8	7.9	
S.E.	27.29	28.34	16.28	36.35	16.06	14.06	20.76	16.05	

Means with similar letters are not significant at 5% level.

186

Table 4: Grain yield of RD 7 (rice variety) with different sources and levels of nitrogen.

Treatment number	Description of treatment	Nitrogen rate by calculation (kg N/ha)	Grain yield (kg/ha)	DMRT
1	Control	0	2234	de
2	Azolla	19	2212	e
3	Azolla	38	2857	abc
4	Azolla	56	2835	abcd
5	Urea	19	2478	cde
6	Urea	38	2516	bcde
7	Urea	56	2555	bcde
8	Urea	75	2537	bcde
9	Azolla + Urea	56 + 19	2926	a
10	Azolla + Urea	19 + 56	2872	ab

C.V. = 8.6%

Means with similar letters are not significant at 5% level.

Table 5: Azolla growth with different type of organic matter as sources of phosphate.

Treatment	Azolla fresh weight (kg/rai)		
	1st Crop	2nd Crop	3rd Crop
Control	487 d	404 b	193 d
Azolla + Superphosphate	1449 a	1331 a	729 c
Azolla + Bran	929 c	1457 a	1621 a
Azolla + Compost	1053 b	1550 a	1251 b

Means with similar letters are not significant at 5% level.

Table 6: Details of treatment of the study on comparison among kinds
of phosphate fertilizer.

Treatment number	Fertilizer rates $N-P_2O_5-K_2O$ (kg/rai)	Sources of phosphate fertilizer
1	No Azolla	0-4-4	Superphosphate 20% P_2O_5
2	No Azolla	6-4-4	Superphosphate 20% P_2O_5
3	Azolla	0-4-4	Superphosphate 20% P_2O_5
4	Azolla	3.2-4-4	16-20-0
5	Azolla	4-4-4	20-20-0
6	Azolla	4-4-4	16-16-8
7	Azolla	1.3-4-4	6-24-24

Table 7: Grain yield of the study on comparison among kinds of
phosphate fertilizer.

Treatment	Grain yield (kg/rai) Sakon Nakhon	Pimai
1 0-4-4 no Azolla	886 b	391 b
2 6-4-4 no Azolla	931 ab	628 a
3 0-4-4 Azolla	1064 ab	590 a
4 16-20-0 Azolla	1038 ab	677 a
5 20-20-0 Azolla	1025 ab	667 a
6 16-16-8 Azolla	1135 ab	704 a
7 6-24-24 Azolla	1191 a	655 a

Means with similar letters are not significant at 5% level.

18. Azolla as a biofertilizer for rice in Sri Lanka

S. A. Kulasooriya, W. K. Hirimburegama
Department of Botany
University of Peradeniya
Sri Lanka

and S. W. Abeysekera
Rice Research Station
Ambalantota, Sri Lanka

Key words   ARA   Azolla pinnata   biomass   fertilizer   rice
Sri Lanka

Summary

The ability of Azolla pinnata to grow and establish itself in mono-culture in rice fields was examined in several locations, falling within different agro-ecological zones of Sri Lanka. Rapid growth of Azolla was observed in all the localities examined. Periodic phosphorus additions and standing water were essential requirements for Azolla growth in rice fields.

The rate of growth of Azolla was measured during two rice growing seasons at Ambalantota, in the low country dry zone where the terrain is undulating to flat, the rice soils are of the low humic gley type and the 75% expectancy value of annual rainfall is below 500 mm. The rice soil pH in this locality is around 6.8, the diurnal light intensity varies from 5 to 130 Klux, with a daily temperature variation of 25 to 37°C. Fresh weight biomass values of Azolla measured over 2 and 3 week periods, during two seasons, gave doubling times of 4.8 and 3.9 days. The corresponding rates of ARA were 2.31×10^{-6} and 2.73×10^{-6} Mol C_2H_4/g (f.w.)/h. These values are approximately equivalent to 230 to 350 g N/ha/h, when converted on a 1:4 (nitrogen fixed: ethylene produced) ratio in relation to the corresponding Azolla biomasses.

Two incorporations of Azolla (grown in dual culture with rice) during a single crop cycle, resulted in grain yields equivalent to fields that received 55 to 84 Kg N/ha of chemical nitrogen fertil-izer. Azolla in dual culture with rice also brought about a 50% reduction in weed growth.

Azolla growth with rice under different planting patterns produced grain yield increases of 13, 22 and 47% respectively, in broadcast seeded, transplanted and avenue planted rice over control plots that received no nitrogen under the same planting patterns.

These results provide encouraging evidence for:

- the ability of Azolla pinnata to grow rapidly in several rice growing areas in Sri Lanka;
- the suitability of Azolla as a biofertilizer for rice, that can replace a substantial amount of nitrogenous fertilizer

1. Introduction

The importance of Azolla specially as a biofertilizer for rice, is well documented[9][11][6][8][10][12][7][14] but its widespread use by rice farmers is more or less confined to Vietnam and China where it has been used traditionally for centuries in rice cultivation. Even in these countries, the production and utilization of Azolla as an integral part of organized rice production is a relatively recent development[6][12]. In many other Asian countries the use of Azolla in rice cultivation is still at a stage of laboratory and field experimentation.

In Sri Lanka, we commenced research on Azolla in 1976. As a prerequisite to its field use ·in predominantly rice growing dry zones of Sri Lanka, we examined the effect of light, temperature and phosphorous on the growth and nitrogen fixation of an indigenous strain of A. pinnata[4]. We found that this strain was capable of tolerating light intensities and temperatures which were reported to be inhibitory to certain temperate species of Azolla[1]. Similar results indicating that A. pinnata is capable of tolerating the climatic conditions in the tropics were reported by Tung and Shen[13]. Encouraged by these results, we tested the ability of certain indigenous and some exotic strains of A. pinnata to colonize rice fields in different localities of Sri Lanka. We further examined the effect of Azolla on rice at two locations, during three rice growing seasons. The results of these investigations are presented and discussed in this paper.

2. Materials and Methods

2.1 Monoculture of Azolla in rice fields. This was done in field plots whose sizes ranged from 5 to 165 m^2. The larger plots were initially partitioned into subplots of smaller area using bamboo poles to prevent the drifting of Azolla inoculum, because turbulence and fragmentation has been shown to inhibit Azolla growth[1]. The field plots were puddled and bunded by 25 cm high levees and a 10 - 15 cm level of standing floodwater was maintained throughout the experimental period. The plots were inoculated with fresh Azolla at the rate of 100 - 200 g/m^2. The inoculum were usually mixed with concentrated super phosphate (CSP) powder and Carbofuran insecticide (3% a.i.) before adding onto the field.

Periodic visual observations together with fresh biomass measurements were done to evaluate the growth of the inoculated Azolla. This type of preliminary screening of field colonization by Azolla was done at Peradeniya, Undugoda, Pannala and Ambalantota; localities that fall within three major agroclimatic zones of Sri Lanka (Figure 1). A more systematic examination of Azolla growth in the field was carried out during the Yala 1980 (April to August - dry season) at the Rice Research Station at Ambalantota. In this region the terrain is undulating to flat, the rice soils are of the low humic gley type, the 75% expectancy of the annual rainfall is below 500 mm, the soil pH is 6.8, the diurnal light intensities vary from 5 to 130 Klux and the daily temperature variation is between 25 - 37°C. The rate of growth and nitrogen fixation of four strains of A. pinnata (named after their original habitats) as the Debokkawa, (Dk), Peradeniya (Pd), Indian (Ind) and Bangkok (Bk) strains were examined in triplicate field plots (1 m x 5 m) with a standing flood-water level of around 10 cm. Each plot was inoculated with fresh Azolla at the rate of 180 g/m^2, mixed with 6 g/Kg (fresh Azolla) of CSP and 1 g/Kg (fresh Azolla) of Carbofuran. Subsequently, CSP powder (1.5 g/m^2) was broadcast over the Azolla every 5 days and carbofuran was added (0.5 g/m^2) at the initial sign of any pest attacks. The fresh weight of Azolla was measured separately in each plot every 3 days, using a 0.5 m^2 quadrat frame. Within 15 days, most of the plots were completely covered by Azolla and it was felt that space limitation could inhibit Azolla growth at this stage.

2.2 Measurement of 'in situ' nitrogenase activity. Nitrogen fixing activity of these 15-day old Azolla covers were evaluated by Acetylene Reduction Activity (ARA) measurements, using plastic 'baby feeding' bottles as the in situ incubation vessels, as described by Kulasooriya et al[5].

A second evaluation of field growth of Azolla was carried out at Ambalantota during Yala 1982. This was done in a larger plot (30 m x 5.5 m) initially subdivided into two equal subplots. An inoculum of 60 g/m^2 of fresh Azolla together with 12 g of CSP was added to each subplot. Subsequently CSP was added at the rate of 1 g/m^2 of Azolla cover every 7 days. The Azolla grew very rapidly and formed a thick cover over the entire area in 21 days. Biomass measurements and ARA estimations were done on the 22nd day. As there were no pest attacks, carbofuran was not added during this period. Biomass of the Azolla cover was measured by taking the fresh weight of 10 random samples, each 1 m^2; ARA measurements were done separately on green and red patches of Azolla. Eight random samples were tested for each patch of Azolla.

2.3 Effect of Azolla on rice under different planting patterns. Transplanting at 20 x 20 cm is recommended for the planting of most high yielding, modern varieties of rice. However, due to limitations of irrigation and water control and to reduce labor cost, rice farmers in many parts of Sri Lanka still adopt direct seeding of rice. Chinese scientists have advocated a change in the planting pattern of rice whereby rice is planted in narrow rows alternating with wider rows to permit a better growth of Azolla in dual culture with rice[6]. An experiment was carried out at Peradeniya to examine the effect of Azolla on broadcast seeded, normally transplanted and avenue planted (15 x 30 x 15) rice.

This experiment was done in 4 m x 3 m plots, replicated three times per treatment. The treatments were, direct seeded rice (125 g/plot), 18-day old seedlings of rice (Variety BG 11-11) transplanted at 20 x 20 cm or avenue planted at 15 x 30 x 15 cm. The planting density of the latter two treatments was 250 hills per plot. Triplicate control plots per treatment had the same planting patterns, but no Azolla was added to them.

All the plots received a basal dressing of 20 Kg N/ha as urea and 37.5 Kg of K_2O/ha as muriate of potash (MOP). The control plots received a basal dressing of 75 Kg/ha of CSP while the Azolla plots received 20 Kg/ha of P_2O_5 as a basal dressing, followed by 55 Kg/ha applied in 6 splits on the Azolla cover as its growth progressed. Azolla was inoculated 7 days after transplanting and 14 days after direct seeding, at the rate of 100 g (f.w.)/m^2. Data was collected with regard to growth (fresh weight) and nitrogenase activity (ARA) of Azolla; growth of weeds (fresh weight), with and without Azolla, and the grain and straw yield of rice under different planting patterns, with and without Azolla.

2.4 Effect of Azolla on rice in comparison to chemical fertilizer. The rapid growth of Azolla in monoculture under field conditions at Ambalantota prompted us to conduct a series of comprehensive field trials to examine the effect of Azolla on rice in comparison to fertilizer nitrogen. This experiment was conducted for 3 consecutive rice growing seasons: Yala 1981 (April–August); Maha 1981–82 (October to February); and Yala 1982.

This experiment was carried out in 4 m x 4 m plots on a randomized-complete-block design, with four replications per treatment. The treatments were:

1. Control (no Azolla, no fertilizer)
2. ON + fertilizer*
3. 2ON + fertilizer
4. 4ON + fertilizer
5. 6ON + fertilizer
6. 8ON + fertilizer
7. 100N + fertilizer
8. Blue-green algae + fertilizer
9. Azolla (Pd) + fertilizer
10. Azolla (Dk) + fertilizer
11. Azolla (Bk) + fertilizer

* CSP and MOP supplied at levels recommended for this variety of rice, applied as basal and at panicle initiation.

A basal dressing of 10 Kg N/ha of urea fertilizer was added to all the treatments except 1 and 2, to ensure the proper establishment and healthy growth of the rice seedlings. Two top dressing of urea were subsequently given to treatments 3, 4, 5, 6 and 7, two weeks and six weeks after transplanting, to provide respectively 55% and 45% of the corresponding N-levels to be supplied. A 3½ month, local variety of high yielding rice (AT-16) was used at a 20 x 20 cm spacing of transplanting. Independent irrigation and drainage was provided to all the plots. Azolla was inoculated to treatments 9, 10 and 11, four days after transplanting, at the rate of 125 g (f.w.)/ m^2, while blue-green algae was inoculated to treatment 8 at the rate of 62.5 g (f.w.)/m^2. These inocula were mixed with 10 g CSP and 5 g Carbofuran before application to the field.

Data was collected with respect to Azolla growth (f.w.), nitrogenase activity (ARA), straw and grain yield of rice and weed growth (f.w.). Azolla was incorporated manually twice during the growth of the crop 4 weeks and 7 weeks after transplanting.

3. Results

3.1 Monoculture of Azolla. Prolific growth and rapid colonization of rice yields by Azolla was observed in all the localities examined, and there were no significant differences in the growth rates of the different strains tested. The pattern of growth exhibited by the 4 strains of Azolla tested at Ambalantota is shown in Figure 2. It can be seen from this figure that all the strains showed a similar of growth with a 7-8 fold increase in biomass in 15 days, giving a doubling time of 4.8 days.

The biomasses of Azolla at full cover in 15 days and their ARA are given in Table 1. It is clear from this table that the biomasses ranging from 7.1 to 8 Kg per plot are equivalent to 14 to 16 t/ha of fresh Azolla showing ARA equivalent to 2.3 to 3.5 Kg N/ha/day when converted on 1:4 N_2 fixed: ethylene produced ratio, and extrapolated on a 10 h daylight period in relation to their biomasses.

The monoculture of Azolla in the 165 m^2 plot during June–July 1982, produced even faster growth. The 10 Kg inoculum multiplied itself 50-fold to yield 511 Kg in 22 days, giving a doubling time of 3.9 days. The final biomass of the standing Azolla cover on day 22 was equivalent to 31 t/ha. The mean ARA values for "green Azolla" in this cover was $3.12 \pm 0.31 \times 10^{-6}$ mol C_2H_4/g (f.w.)/h, while that for the "red Azolla" was 2.35 ± 0.29. This difference in ARA between the green and red Azolla is not statistically significant at the 5% level. These ARA values when converted as described earlier are equivalent to 6.8 and 5.1 Kg N/ha/day for the green and the red Azolla respectively.

3.2 Effect of Azolla on rice under different planting patterns. The fresh weights of Azolla and their corresponding ARA under different planting patterns of rice are given in Table 2. These Azolla biomasses given in the table as g (f.w.)/m^2 when extrapolated on the presumption that half the total field area is occupied by the rice hills, gives values of 6.5, 11.5 and 15 t/ha of Azolla under broadcast seeded, transplanted and avenue planted rice respectively. The ARA values converted on a 1:4 ratio and extrapolated on the same basis shows potential contributions of 0.63, 1.57 and 2.04 Kg N/ha/day under the corresponding planting patterns. The incorporation of these biomasses resulted in grain yield increases of 14.22 and 47% respectively under the different planting pattern (Table 3).

3.3 Effect of Azolla on rice in comparison to chemical fertilizer. Diurnal ARA of Azolla in dual culture with rice, 15 days after inoculation is shown in Figure 3. It is seen from this figure that all three Azolla strains exhibited noteworthy nitrogenase activity by 900 h and these values reached a maximum aroudn 1200 h, when the light intensity was 100 – 125 Klux and the temperature was 33 – 34°C. Even at 1800 h when the light intensity was only 5-7 Klux, there was still some detectable activity.

Fresh biomasses of the full cover of Azolla produced in 27 days under dual culture with transplanted rice (Table 4) give values ranging from 2.0 to 2.2 Kg/m^2 which are equivalent to 10 to 11 t/ha. It can also be seen that the biomass of Azolla (Peradeniya strain) produced under dense planting was not significantly different from that under 20 x 20 cm transplanted rice, and that avenue planting has permitted only a marginal increase.

These results show that all the three stains of Azolla could grow harmoniously with rice and fix nitrogen very well in this dry zone habitat.

The grain yield data obtained with 2 Azolla incorporations during the crop cycle, in comparison to different levels urea N-fertilizer treatments are in Table 5. This table shows that 3.74 to 3.98 t/ha of brown rice could be produced in these fields without any added fertilizers and that such yields could be increased to 3.99 and 4.31 t/ha by P and K supplementation. There is also a clear, positive yield response to chemical N-fertilizer additions. The Azolla incorporations have resulted in grain yield increases of 14 to 39% over the control (ON + F), whereas the BGA treatment produced only a 3% increase, which is not significantly different at the 5% level. The low yield response to Azolla during Yala 1982 must be viewed in conjunction with similar low responses even to chemical N-additions, where even at the 100 Kg N level the increase in yield over the control was only 23%.

Linear correlations between grain yields and fertilizer N-additions during the 3 seasons are shown in Figure 4. Reading from these curves, it is seen that Azolla incorporations could produce grain yields equivalent to plots that would have received 55 to 84 Kg N/ha of chemical fertilizer as urea.

Nevertheless, it would not be correct to interpret the beneficial role of Azolla purely in terms of the nitrogen input. Besides increasing soil fertility, a good cover of Azolla could suppress weed growth. The results in Table 6 shows that Azolla suppressed weed growth by 34 to 53% in our experiments under the different planting patterns.

4. Discussion

The ability of Azolla to grow in rice fields in a number of locations, in different agro-climatic zones of Sri Lanka indicates that this plant has a good chance of successfully colonizing rice fields under our tropical natural conditions.

All the four strains of A. pinnata tested at Ambalantota showed a similar pattern of growth in this dry zone (Figure 2) indicating that they are well adapted to the highlight and temperature conditions in this region. Similar results reported for the same species of Azolla in Malaysia (Tung & Shen 1981) shows that this species, indigenous to South East and South Asia is better suited for use in tropical habitats than temperate species like A. filiculoides which are susceptible for such conditions .

The doubling times of 4.9 and 3.8 days for the field growth of Azolla fall within the range reported by Becking[2].

The nitrogen fixing rates calculated from in situ ARA measurements are rather high, perhaps due to extrapolation of short term estimations, done under optimal conditions employing a theoretical relationship of N fixed to ARA. Nevertheless these values reflect a maximum potential available for exploitation.

The biomasses of Azolla of 30 t/ha in monoculture and 10 t/ha in dual culture with rice, have the potentials to provide 75 and 25 Kg N/ha respectively. These values are compatible with the range reportd by Kikuchi et al[3].

Our results show that Azolla inocula of 1.8 and 0.6 t/ha added to rice fields during Yala 1981 and Yala 1982 were capable of forming full covers in 2 weeks and 3 weeks. The better growth in 1982 was due to more cloudy days during the latter season, which shows that

even these strains of A. pinnata grow better under less light than
under full sunlight, whose intensity remains about 130 Klux for
2 to 4 hours around noon time, in this region. These results also
show that it is possible to raise an Azolla crop after land preparation so that a basal dressing of organic nitrogen may be provided
by the incorporation of Azolla prior to transplanting of rice. Such
Azolla production however needs a supply of standing water as an
essential requirement. This in fact is a severe constraint in many
parts of Sri Lanka, where rice is grown under rainfed conditions.
Even under irrigation, water supply is controlled and an uninterrupted
supply 3 weeks ahead of schedule is not frequently provided. It
is primarily because of this that we did not attempt to grow Azolla
prior to rice transplanting in our Azolla-rice dual culture experiments. The dual culture of Azolla under different planting patterns,
at Peradeniya has shown that Azolla incorporation resulted in 14,
22 and 47% increases in grain yield under broadcast seeded, transplanted and avenue planted rice respectively. These yields obtained
in a relatively infertile field which could produce only 1.44 t/ha
of grain in the control plots, showed that the change in planting
patterns had a positive effect on Azolla, which enhanced the grain
yield. However, in a similar experiment conducted at Ambalantota
on more fertile soils (which produced more than 3 t/ha in the control
plots), Azolla growth under transplanted and avenue planted rice
was not significantly different (Table 4), and the growth of Azolla
(Pd) under transplanted and densely random planted rice was also
not significantly different at 5% level. These results perhaps
indicate that altering the planting pattern of the crop may not
bring about significant increases in Azolla growth in fertile fields.

The series of experiments conducted for three consecutive seasons
at Ambalantota has shown that Azolla use can minimize fertilizer
N-additions quite substantially in this region of Sri Lanka. From
Figure 4 it is clearly seen that two incorporations of Azolla could
produce grain yields equivalent to fields that received 55 to 84
Kg N/ha. These values indicate that Azolla used as a biofertilizer
has the potential to replace 120 to 183 Kg/ha of urea fertilizer.
Such amounts come very close to the fertilizer levels recommended
for this area and are in excess of the actual fertilizer levels
applied by the peasant farmers.

Although it is customary to attribute such grain yield increases
to nitrogen fixation by Azolla, this is not an accurate interpretation. It can be seen from Table 5 that a good cover of Azolla has
the ability to smother weed growth and bring about a 50% reduction
of weed biomass.

An Azolla incorporation adds a considerable amount of organic
matter to the soil and this should invariably improve soil texture,
soil aeration, cation exchange capacity etc. Although we have no
field data on these parameters, it is very likely that the beneficial
effects of Azolla on the crop is at least partly due to physical
improvement of the soil.

Taken together, these results clearly show that Azolla is an attrac-
tive alternative to costly chemical fertilizer nitrogen, for rice
production in Sri Lanka. However, more extensive research, preferably
in farmers fields, are necessary especially to evaluate the cost:
benefit ratio in relation to water supply and labor, before Azolla
technology is recommended for widespread application.

Acknowledgements   We thank Professor M. D. Dassanayake, Head of Department of Botany, University of Peradeniya and the Director of Agriculture for his constant encouragement received during these investigations.   These research projects were supported by funds received from the National Science Council of Sri Lanka and the International Foundation for Science (IFS), Sweden.   The senior author (S.A.K.) gratefully acknowledges the travel grant awarded by the IFS which enabled his participation at this workshop.

References

1 Ashton P J   1974   The effect of some environmental factors on the growth of Azolla filiculoides Lam. In: The Orange River Pro- gress Report. (Institute for Environmental Sciences), University of Bloemfontein, South Africa. pp. 123–138.

2 Becking J H   1979   Environmental requirements of Azolla for use in tropical rice production.   In: Nitrogen and Rice. The Inter- national Rice Research Institute, Los Banos, Philipppines. pp. 345–374.

3 Kikuchi M, Watanabe I and Haws L D   1982   Economic evaluation of Azolla use in rice production.   Int. Conf. on Organic Matter Production. The Interrational Rice Research Institute, Los Banos, Philippines (In Press).

4 Kulasooriya S A, Hirimburegama W K and de Silva R S Y   1980 Effect of light, temperature and phosphorus on the growth and nitrogen fixation in Azolla pinnata, native to Sri Lanka.   Oecol. Plant 355–365.

5 Kulasooriya S A, Hirimburegama W K and Abeysekera S W   1982 Growth and nitrogen fixation in Azolla pinnata under field conditions. Jour. Natn. Sci. Council of Sri Lanka (In Press).

6 Liu Chung Chu 1979   Use of Azolla in rice production in China. In: Nitrogen and Rice. The International Rice Research Institute, Los Banos, Philippines. pp 375–394.

7 Lumpkin T A and Plucknett D L   1980   Azolla: botany, physiology and use as green manure. Econ. Bot. 34: 111–153.

8 Rains D W and Talley S N   1979   Use of Azolla in North America. In: Nitrogen and Rice. The International Rice Research Institute, Los Banos, Philippines. pp. 419–434.

9 Moore A W   1969   Azolla: biology and agronomic significance. Bot. Rev. 35: 17–35.

10 Singh P K   1979   Use of Azolla in rice production in India. In: Nitrogen and Rice. The International Rice Research Institute, Los Banos, Philippines. pp. 407–418.

11 Talley S N, Talley B J and Rains D W   1977   Nitrogen fixation by Azolla in rice fields.   In: A. Hollaender (ed.). Genetic Engineering for Nitrogen Fixation. Plenum Press, N.Y. and London. pp. 259–281.

12 Tuan D T and Tran Q T   1979   Use of Azolla in rice production in Vietnam.   In: Nitrogen and Rice. The International Rice Research Institute, Los Banos, Philippines. pp. 395–406.

13 Tung H F and Shen T C   1981   Studies on the Azolla pinnata-Anabaena azollae symbiosis: growth and nitrogen fixation. New Phytol. 87: 743–749.

14 Watanabe I, Bai K Z, Berja N S, Espinas C R, Ito O and Subudhi B P R   1981   The Azolla-Anabaena complex and its use in rice

culture.   Int. Rice Res. Paper Series 69. 10 pp.

Table 1: Biomass and Acetylene Reducing Activity (ARA) of 15-day old monoculture of <u>Azolla pinnata</u> strains grown in 5 m^2 field plots at Ambalantota, in the low-country, dry zone of Sri Lanka. Growth conditions are the same as in Figure 2.

Azolla strain	Fresh weight of Azolla[a] (g/plot)	ARA[b] 10^{-6} mol C_2H_4/g (f.w.)/h	N$_2$ Fixation (kg N/ha/day)[c]
Debokkawa	8000±54	2.59±1.50	3.48
Bangkok	7892±72	2.44±1.36	3.22
India	7600±124	1.82±1.18	2.32
Peradeniya	7125±712	2.41±1.36	2.88

[a] Mean value of four replicates.
[b] Mean value of 8 samples incubated with 20% acetylene from 1330 to 1430 CST, under 90 klux at 34 to 37°C.
[c] N$_2$: C$_2$H$_4$ = 1:4

Table 2: Fresh weight of <u>Azolla</u> and its Acetylene Reducing Activity (ARA) under rice, 27 days after inoculation.

Treatment	Fresh weight of Azolla (kg m^{-2})	ARA (mmol C_2H_4 h^{-1} m^{-2})	N$_2$ Fixation[a] (kg N$_2$ ha^{-1} day^{-1})
Broadcast seeded	1.34	1.81±0.08	0.63
Transplanted at 20 cm x 20 cm	2.33	4.50±0.33	1.57
Transplanted in avenues	2.99	5.83±0.31	2.04

[a] N$_2$: C$_2$H$_4$ = 1:4

Table 3: Grain yield and straw yield of rice grown under three planting patterns, with and without Azolla.

Treatment		Grain yield		Straw yield	
Planting pattern	Azolla	t ha^{-1}	% increase	t ha^{-1}	% increase
Broadcast seeded	–	1.44		2.77	
	+	1.64	14%	3.45	25%
Transplanted at 20 cm x 20 cm	–	2.25		2.56	
	+	2.74	22%	3.10	21%
Transplanted in avenues	–	2.07		2.47	
	+	3.04	47%	3.41	39%
		cv = 2.86%		cv = 13.5%	

Table 4: Maximum biomass produced in 27 days by Azolla in dual culture with rice at Ambalantota.

Azolla strain	Transplanting pattern of rice	Biomass* (g m^{-2})
Debokkawa	20 cm x 20 cm	2200 ab
Bangkok	20 cm x 20 cm	2190 ab
Peradeniya	20 cm x 20 cm	2070 c
Peradeniya	Avenue	2350 a
Peradeniya	Dense	1970 c

* Any two means followed by the same letter are not significantly different at 5% level.

Table 5: Effect of Azolla and Blue Green Algae (BGA) on Rice

(dual culture and two incorporations of Azolla after transplanting of rice)
Grain yield data for 3 consecutive seasons (1981-82)
Rice variety - At 16 (3½ months)

Treatment	Grain yield (t ha^{-1}) Season			Percentage increase over F.0N		
	1981 Yala	1981-82 Maha	1982 Yala	1981 Yala	1981-82 Maha	1982 Yala
Rice only	3.74 a	3.89 a	3.98 a	–	–	–
F. 0N	3.99 b	4.16 ab	4.31 b	–	–	–
F. 20N	4.55 cd	4.39 b	4.54 bc	14%	5%	5%
F. 40N	4.79 d	4.59 b	4.63 bcd	20%	10%	7%
F. 60N	5.09 e	5.38 c	4.87 cde	27%	29%	13%
F. 80N	5.27 f	5.50 c	5.04 ef	32%	32%	17%
F. 100N	5.56 g	5.91 c	5.29 f	39%	42%	23%
F. 10N. BGA	4.11 b	3.94 ab	–	3%	–	–
F. 10N. Azolla-Pd	5.39 fg	5.81 c	4.92 de	35%	39%	14%
F. 10N. Azolla-Dk	5.46 fg	–	–	37%	–	–
F. 10N. Azolla-Bk	5.26 ef	–	–	32%	–	–
	cv = 6.02%	cv = 9.2%	cv = 4.4%			

[1] Yala season = April to August
Maha season = October to February
Treatments with same letter are not significantly different.

Table 6: Dry weight of weeds at 30 days after transplanting of rice, with and without Azolla.

Planting pattern	Azolla	Peradeniya experiment		Ambalantota experiment	
		g.plot⁻¹	% suppression	g.plot⁻¹	% suppression
Broadcast seeded	−	35.0		n.d.	n.d.
	+	23.3	34.%	n.d.	
Random transplanting	−	n.d.		383	
	+			180	53%
Avenue transplanting	−	100.6		510	
	+	55.6	45%	297	42%
20 cm x 20 cm transplanting	−	74.3		699	
	+	48.6	35%	331	52%

n.d. = not determined
Treatments with same letter are not significantly different.

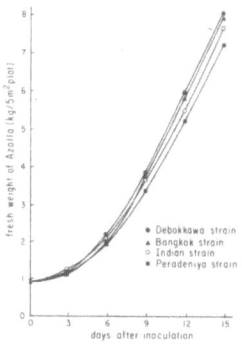

Figure 1. Map of Sri Lanka showing the major climatic zones and the sites where investigations with Azolla were carried out.

Figure 2. Growth patterns of different strains of Azolla pinnata in 5 m² field plots at Ambalantota. Each plot was initially inoculated with 900 g (f.w.) of Azolla together with 6 g/kg (fresh Azolla) of CSP and 1 g/kg (fresh Azolla) of Carbofuran (3% a.i.). Subsequently CSP powder (1.5 g/m²) was broadcast over the Azolla cover every 5 days of Carbofuran (0.5 g/m²) was added at the initial sign of any pest attacks. (Diurnal light intensity; 5 to 125 klux; daily temperature: 25 to 37°C).

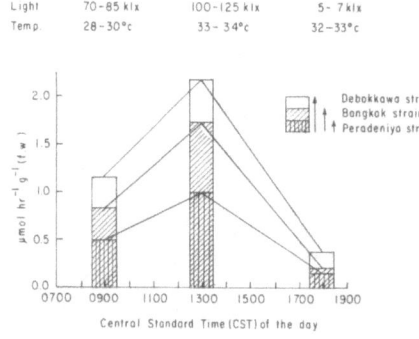

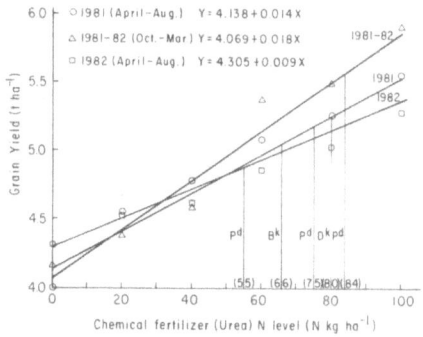

Figure 3. Diurnal variation of Acetylene Reduction Activity (ARA) of three Azolla isolates under field conditions in dual culture with rice.

Figure 4. Estimated linear relationship between grain yield and chemical fertilizer (Urea) N levels for three consecutive seasons. Figures in parenthesis indicate the performance of the Azolla incorporated treatments in terms of kg N in the form of Urea fertilizer.

19. Azolla in West Africa. First results from the WARDA program

H. F. Diara
Association pour le Dévelopement de la Riziculture en
Afrique de l'Ouest, B.P. 29, Richard Toll, Sénégal

C. A. Dixon
West Africa Rice Development Association
P. O. Box 7
Rokupr, Sierra Leone

C. Van Hove
Université Catholique de Louvain,
Place Croix du Sud, 4 B-1348
Louvain-la-Neuve, Belgium

Key words   Azolla   Green manure   Nitrogen   Rice

Summary

Introduction of Azolla as green manure for rice has been tested
in two ecologically very different agronomic stations from the West
Africa Rice Development Association (WARDA): Fanaye, North Senegal,
with a semi-arid climate, and Rokupr, North Sierre Leone, with a
tropical rainy climate of the mangrove type. First results, which
appear promising, are presented.

1. Introduction

The West Africa Rice Development Association (WARDA) is an inter-
governmental organization whose objective is to help the fifteen
country members, representing some 150 million people, to become
self-sufficient as regards rice production; in 1980 the rice self-
sufficiency rate was 47.7%.
   Among the numerous limiting factors of productivity in the region,
fertilizer, and particularly lack of available nitrogen fertilizer,
is of increasing importance.
   Two years ago WARDA developed a research program on Azolla designed
to reduce chemical nitrogen fertilizer consumption.
   Taking into account the great climatic and pedological diversities
encountered in West Africa, two WARDA pilot agronomic stations were
selected: Richard Toll - Fanaye, North Senegal, with a semi-arid
climate, where water supply is provided all year by pumping, allowing
2 or even 3 annual crops, and Rokupr, North Sierra Leone, with a
tropical rainy climate, where water control and salinity are worrying
problems, trials being done at the limit of cleared mangrove swamps.
Climatic data for these two stations are presented on Table 1 and
Figures 1 and 2.

2. Materials

Azolla being not naturally present in north Senegal, twenty-four strains were introduced at the Richard Toll research center and after a first screening of their behavior, A. pinnata strain UCL-7 (Indian origin) was selected for preliminary field trials.

At Rokupr a local strain has been selected (A. pinnata strain UCL-38). Rice variety at Fanaye: I KONG PAO; at Rokupr: ROK 5.

3. Methods

3.1 First trial at Fanaye. Experimental design: randomized blocks; spacing: 25 x 25 cm; plots: 4.5 x 4 m; insecticide: Furadan: 1 kg a.m./ha (three applications); fertilization: N= urea (cfr. treatments) or Azolla; P= 60 kg P_2O_5 /ha; K= 25 kg K_2O/ha; treatments: cfr. table 2; replications: 5.

3.2 INSFFER trials at Fanaye. General conditions are similar to those concerning the first trial, but for:

- spacing: 20 x 20 and 40 x 10 cm.
- P fertilization: 60 kg P_2O_5/ha on the soil. For treatments without Azolla 45 kg P_2O_5/ha on the soil, 15 kg pulverized for treatments with Azolla.
- Treatments: cfr. Table 3

3.3 First trial at Rokupr. Experimental design: randomized blocks; spacing: 25 x 25 cm; plots 4 x 3.5 m; Furadan: 1 kg a.m./ha; replications: 5.
Fertilization: N: 40 kg N/ha as urea injected at 15 - 20 cm depth (locally recommended dose) or Azolla; P: 21 kg P_2O_5/ha.
Treatments: Az. I: 17   t   Azolla/ha,   before   rice   transplantation
              Az. II: 17 t Azolla/ha, two weeks after transplantation.

3.4 Parameters. For all these experiments paddy yield (14% humidity), straw yield, panicle density and plant height at maturity have been recorded. Description of results which follows only takes into account the first and most important of these parameters.

3.5 Statistical analysis. Duncan's multiple range test has been used for the first Fanaye trial, Newman-Keuls test for the other experiments.

4. Results and Discussion

4.1 First trial at Fanaye (Table 2). Treatments 2, 4 and 7 included Azolla inoculation after rice transplantation, the purpose being to test the effect of intercropped Azolla on the following rice culture. Nevertheless Azolla developed together with rice not only in these plots but in all the others, due to the action of birds and frogs and to an incomplete soil incorporation of Azolla. As a result treatments 9 and 4, as well as 8 and 7, have to be considered as practically similar; further, treatment 4 low yields are attributable to a particularly poor incorporation of Azolla, due to an imperfect drainage before ploughing in two replications from these treatments.

These unanticipated difficulties notwithstanding, these first results clearly suggest the potential value of Azolla as nitrogen fertilizer under the sahelian conditions.

4.2 INSFFER trials at Fanaye (Table 3). The INSFFER protocol has been followed successively during the rainy season 1981, the cold season 1981-82 and the dry, hot season 1982. The locally recommended level of N fertilizer being 120 units (as urea), this treatment has been added.
The main conclusions of these experiments are:

- all the treatments were always significantly more productive than the control; their effects, as related to it, increased with time.
- the recommended N dose generally gave the highest yields.
- incorporation of three Azolla mats, or one Azolla mat plus 30 N units gave yields comparable to those obtained with 60 N units.
- Results concerning spacing effects do not allow clear conclusions.

4.3 First trial at Rokpur (Table 4). From this preliminary experiment it may be concluded that incorporation of one Azolla mat (± 17 t. F.W.) before rice transplantation significantly increased yields, while a second incorporation, two weeks after rice transplantation gave yields similar to the locally recommended N dose at the time of experiment, (this dose having increased to 60 kg N since then). Water control seems to be the most worrying problem in the area tested.

Acknowledgements   Thanks are due to Dr. H. K. Pande, CRRI, Cuttack (India), for having kindly provided one of the Azolla strains.

Table 1: Climatic data for Fanaye and Rokupr

	Fanaye Semi-arid climate	Rokupr Tropical rainy climate
Longitude	15° – 42' W	12° – 57' W
Latitude	16° – 27' N	09° – 01' N
Rainfall	300	3000
Monthly mean daily hours of bright sunshine	11.3 – 13.0	2.1 – 8.1
Monthly maximum mean temperatures	30 – 41°C	27 – 34°C
Monthly minimum mean temperatures	14 – 25°C	20 – 23°C

Table 2: Rice yield as affected by urea and/or Azolla
(Fanaye, 14/03 to 25/06/81)

Treatment (1)	Paddy yield	% Control	Duncan's multiple range test 5% (2)	
1 Control	2184	100	a	
2 60 + Azolla 3	2349	108	a	
3 Azolla 1	2634	121	a	b
4 Azolla 2+Azolla 3	2976	136	a	b
5 Azolla 2+30 b	3107	142		b
6 Azolla 2+30a+30b	3118	143		b
7 Azolla 2+Azolla 3+30A	3128	143		b
8 Azolla 2+30A	3317	152		b
9 Azolla 2	3382	155		b
10 60+30a30b	3413	156		b

(1) 60 = 60 kg N before rice transplantation
30a = 30 kg N at tillering
30b = 30 kg N at earing
Azolla 1 = 1 Azolla mat ploughed in before rice transplantation
Azolla 2 = 2 Azolla mats ploughed in before rice transplantation
Azolla 3 = Azolla cultivated with rice.
(2) Treatments with same letter are not significantly different at level stated.

Table 3: INSFFER trials - Fanaye paddy production (% control).

Treatment (1)		Rainy Season (02/09 - 10/12/1981)		Cold Season (24/12 - 13/05/1982)		Hot Dry Season (03/03 - 30/06/1982)	
Control	20 x 20	100 (3300 kg/ha)	a	100 (2324 kg/ha)	a	100 (1699 kg/ha)	a
60	20 x 20	165	b c	180	b c d	318	b
60	40 x 10	152	b	243	e	303	b
30 + Azolla 1	20 x 20	158	b	142	d	331	b
30 + Azolla 1	40 x 10	151	b	210	b e	282	b
Azolla 1 + Azolla 3	20 x 20	152	b	150	c d	312	b
Azolla 1 + Azolla 3	40 x 10	154	b	191	b c	314	b
120	20 x 20	183	c	254	e	448	c
120	40 x 10	158	b	323	f	454	c

(1) 60 = 60 kg N (1/3 + 1/3 + 1/3)
    30 = 30 kg N (1/3 + 1/3 + 1/3)
Azolla 1 = 1 Azolla mat ploughed in before rice transplantation
Azolla 3 = idem Azolla 1 + 2 Azolla mats after transplantation
Treatments with same letter are not significantly different.

Table 4: Azolla use on the limit of cleared mangrove swamp.
(Rokupr, 15/08/81 to 20/11/81).

Treatment (1)	Paddy production	
	kg/ha	% control
Control	2531 a	100
40	3652 b	144
Azolla 1	3164 b	125
Azola 1 + Azolla 2	3581 b	141

(1) 40 = 40 kg N as urea injected at 15 cm depth

Azolla 1 = 17 t Azolla/ha, before rice transplantation
Azolla 2 = 17 t Azolla/ha, two weeks after rice transplantation.
Treatments with same letter are not significantly different.

FOURTH SECTION: ABSTRACTS

20. Response of rice to nitrogen fertilization in Puerto Rico

Fernando Abruña
Agricultural Experiment Station, USDA/ARS
College of Agricultural Sciences
Mayaguez, Puerto Rico

Abstract

In Puerto Rico, nitrogen is by far the most important element in rice production. However, responses are oftenly modified by methods of application, water management systems, season of the year, straw management and pest control. Splitting the N applications increased the efficiency as compared to applying all at planting time. Rice yields were increased steadily from 0 to 224 kg of N/ha attaining a maximum yield of about 8 t/ha when all N was applied at once, whereas similar yields were obtained with only 112 kg of N/ha when N was split with no further increase with heavier applications. Residual effect of N was more evident when fertilizer to previous crop was split applying half at planting time and half at panicle initiation. Higher yields and a more efficient use of N is to be expected when rice is planted from January to June. During this season, a maximum yield of 6.7 t/ha of rough rice was obtained with 168 kg of N/ha, whereas when planted from July to November rice barely yielded 5 t/ha and responded to only 112 kg of N/ha. This reduction in yields was probably due to shorter, cloudy days during the rainy season from July to November. Damage by blast which increased also during the season was correlated with N applications. The heavier the N application, the more severe the disease damage. Disking in the straw after cropping, produced somewhat higher yields at comparable N rates than burning or throwing it away. The difference was more evident after the 5th consecutive rice crop.

212

21. Research on the mass culture of <u>Azolla caroliniana</u> and <u>A.</u> <u>filiculoides</u> in Italy

M. C. Margheri, L. Tomaselli, C. Filpi
Istituto di Microbiologia Agraria e Tecnica dell'Universita
degli Studi e Centro di Studio dei Microorganismi Autotrofi del
C.N.R. - Firenze (Italy)

Abstract

The mass culture of <u>Azolla</u>, since long time practiced in China, is gaining interest in the western countries, because this fern can contribute to reduce the high synthetic nitrogen fertilizer consumption through its efficient symbiotic nitrogen-fixing system.
Hence, in this research Centre since 1977 an investigation has been undertaken concerning clone selection of <u>Azolla caroliniana</u> and <u>A. filiculoides</u>, the physiology of the <u>Azolla-Anabaena</u> symbiosis and the definition of the conditions for laboratory culture as well as for outdoor mass culture. The first outdoor culture of <u>A.</u> <u>caroliniana</u> on synthetic medium utilizing ponds of 1-25 m^2, gave yields of 3-4 d.w./m^2 day^{-1} during the four most favorable months and 2-3 g during the remaining 100 days on which outdoor culture is practicable under Florence's climatic conditions (Margheri et al., 1978). In the last two years particular attention has been paid to maintaining optimum densities of plants (35-40 g d.w./m^2). With this improvement in the culture technique, mean yields of 5-6 g d.w./m^2 per day (maximum of 7.8 g/m^2 day^{-1} for short periods) were attained with <u>A. filiculoides</u>.
The possibility of culturing <u>Azolla</u> in biologically treated waste waters has been tested. In these experiments sugar refinery waste, previously treated with photosynthetic bacteria, were used. This waste water has a low content of combined nitrogen and is relatively rich in K, Mg, P and Fe. In such water (BODs = 300-400 ppm) the yield of <u>Azolla</u> culture were slightly higher, compared with the culture in synthetic medium.
The results obtained till now show that <u>Azolla</u> culture, either in synthetic medium or in waste water, give an average gain in fixed nitrogen of 350 kg/ha during the period May-September. So we may assume that the mass culture of <u>Azolla</u> is convenient, since it realizes a nitrogen enrichment equivalent to the requirement of an agricultural crop.
Another line of investigation concerns the utilization of <u>Azolla</u> as an inoculant for paddy soils. Field experiments carried out with <u>A. filiculoides</u> in rice fields of Oltrepo Pavese has evidenced the attitude of this species to colonize rapidly the rice field giving positive effects on the weed control and on the resistance of the rice plants against flattening. <u>Azolla</u> cultivation in paddy fields can be considered as a new and efficient microbiological green-manuring practice, yielding considerable amount of fixed nitrogen, easily degradable organic matter and an humussium by <u>Azolla</u> makes its biomass slowly mobilized source of this element (Chungchu et al., 1982). Up-to-date conditions and time for a rational

introduction of <u>Azolla</u> in the rice production cycle are to be better defined.   Research in progress are aimed to a comparative study of the physiological and cultural properties of different species and strains of <u>Azolla</u>, to the isolation and symbiotic N_2-fixing cyanobacteria and to a better knowledge of their symbiotic and N_2-fixing properties.

22. Practical utility of Azolla in Samba season in Thanjavur Delta

S. Srinivasan
Blue Green Algae Scheme
Tamil Nadu Rice Research Institute
Aduthurai-612 101
Tamil Nadu, India

Abstract

Having tried all possibilities of using Azolla in the three cropping
seasons of Thanjavur Delta (double crop wetlands-first crop (Kuruvai)
short duration (June – October); second crop (Thaladi); medium duration (October – February) and single crop wetlands (Samba); long
duration (August – February), it was found to be useful in the
Thaladi and Samba seasons. Of these two seasons, intermittent or
sometimes continuous heavy rains and floods in the Thaladi season
made it impossible to use Azolla. Hence Azolla was tried in Samba
season in 10 selected villages, during 1981–82. In each village
2 farm holdings were fixed. In one hectare of the land fixed for
the trial Azolla was introduced at 1 mt/ha on receipt of water,
keeping half of the area (5000 m^2) as control. As per the finding
of the Author-fresh channel water was let into the field once in
3–4 days after draining the old water so that Azolla could be raised
without 'P' manuring. In 7 villages four layers of Azolla were
incorporated (2 layers before and 2 layers after planting) and in
the rest of the three villages 3 layers were incorporated (one layer
before and 2 layers after planting). In both the treatments P and
K were tried at 50 kg/ha. 'N' was varied (0, 25, 50, 75 and 100
kg/ha) in the control plot, while no 'N' was applied in the Azolla
treatment. The variety tried was CR 1009. The yield recorded showed
that in place where Azolla was incorporated four times the mean
yield was 6847 kg/ha as against 6365 kg/ha for 100 kg N/ha. Three
layers of Azolla incorporation recorded 6295 kg/ha. Thus Azolla
can be a substitute to 'N' fertilizer for samba crop is Thanjavur
Delta when water was plentiful.

23. Azolla for rice production in tropical Africa

Ton That Trinh
Agricultural Office
Plant Production and Protection Division
F.A.O. - Rome

Abstract

Traditional rice farming in tropical Africa is mainly dryland rice.
There are only less than 3 millions hectares of lowland rice in
tropical Africa, some of which consist of deep water rice. But
many countries are exploiting the immense potentiality of lowland
rain feed rice (estimated, by FAO to 150 million hectares of suitable
soils), in order to meet their increasing rice demand and to substitute
the permanent rice based cropping systems on lowland to their shifting
cultivation of upland crops. Since traditional rice farming are
poor small farmers in isolated areas, Azolla as nitrogen supplier
could replace appropriately nitrogen fertilizer in tropical region.
Rice yield in lowland tropical Africa is still very low and 40 to
60 kg/ha of nitrogen are enough to give 3 to 4 tons of rice per
hectare with improved varieties. Such amount could be well supplied
by appropriate Azolla strains in one or two seedings before or
after transplanting. Unfortunately African farmers and extensionists
largely ignore Azolla even if one often can find Azolla in the
vicinity of the rice fields, such as the Bumba in Zaire, Tchibanga
in Congo, Majunga in Madagascar, Mange in Sierra Leone, Vallee du
Kou in Upper Volta, etc.
To introduce the use of Azolla in rice cultivation in Africa,
FAO in its field projects is cooperating with member countries at
on farm trials in five countries with Azolla combining or not with
low dose of nitrogen and phosphate fertilizers. These trials would
fit into various rice ecologies of tropical Africa. In mangrove
rice ecology, it would require Azolla strains tolerant to salinity,
iron and aluminum toxicity during the dry season. For such ecology,
research on Azolla tolerating brackish water (African rice farmers
reintroduced brackish water during the dry season to prevent
thioscydation of the mangrove trial) carried out at the Hebrew
University-Israel is very beneficial. In the "inland swamp" rice
ecology, phosphorus deficiency and iron toxicity are frequently
the main nutrient constraints in the rice production, especially
in the moist forest zone. In East Africa rice is often cultivated
in high altitude (Madagascar alone has 700.000 ha of such rice fields)
and is confronted with cold problem.
Azolla filiculoides, Azolla nilotica mentioned by the Naples
University of Italy, or some strains of Azolla caroliniana mentioned
by the Microbiology Society of Egypt are better suited to this rice
ecology than Azolla pinnata. In the irrigated rice of Sahelian
zone, semi-arid conditions require a better method of conservation
of Azolla in the farmers fields during the long dry season which
is also too cool for the growth of the local Azolla africana. The
creation of fish ponds on the bottom lands in these areas should

be combined with conservation ponds for <u>Azolla</u> that could provide
feed to some species, Tilapia for example.

One should not neglect the role of suppressing long and repeated
weedings with early <u>Azolla</u> inoculation, since weed control is also
one of the main constraints in rice production on small African
farms.

24. Evaluation of the Azolla-Anabaena association effect on the dry matter production and total N of Setaria sphacelata cv. Kazungula

María José Valarini, Paulo Bardauil Alcantara
Researchers of Instituto de Zootecnia
Nova Odessa - SP, Brasil

and Mirna Adamoli de Barros
Probationer from Instituto de Zootecnia
Nova Odessa - SP, Brasil

Abstract

Because of the limitation of the natural N fertilizers supply, new approaches for obtaining N_2 biologically have been explored. This include also the cyanobacteria which not only fixes the N_2 from the air but also synthesizes its own tissue from CO_2 and water. From this point of view, Pteridophytes of the genus Azolla have intrinsic interest because they contain in their cavities Anabaena azollae, a diazotrophic cyanobacterium.
  With the aim to make organic fertilization in humid soils and using the Azolla-Anabaena association, a randomized experiment was done in a greenhouse containing 40 plots with hydromorphic soil cultivated with Setaria sphacelata cv. Kazungula. The work has been carried out in Nova Odessa/SP and the statistical analysis has showed significant differences in dry matter productions and total N for the two first cuts to the treatment with Azolla. It was observed that Azolla cultivation was easily done since the average time for fresh matter duplication about 50 hours. The Azolla chemical analysis showed in the dry matter average percentages of 2.66% N; 0.76% Ca and 0.43% P. These results indicate a good effect of Azolla-Anabaena association as a source of nitrogen in the soil-plant system.

25. Phosphorus nutrition in Azolla

C. S. Weeraratna
Faculty of Agriculture
Ruhuna University College
Matara, Sri Lanka

Abstract

The response to basal application of superphosphate by two strains of Azolla pinnata (Bangkok and Peradeniya strains) under rice-field conditions was studied in two different soils, a Tropaqualf and a Tropudult.

Both strains responded to P application. The Bangkok strain showed higher vegetative growth and contained more N than the Peradeniya strain at any given rate of P application. But, at zero P values the Peradeniya strain fared better. The highest rate of growth in the Bangkok strain growing in Tropaqualf was noted when P applied was equivalent 30 kg P_2O_5 per hectare. The corresponding rate for Peradeniya strain was 45 kg P_2O_5 per hectare. The N contents were also highest at these rates. For Azolla growing in Tropudult, the rates of P_2O_5 necessary to produce highest growth rate in the Bangkok and Peradeniya strains were 45 and 60 kg per hectare respectively. Increasing the rates of P application beyond the values indicated did not have any beneficial effect. Liming the two soils at the rate of 2 tons per hectare, reduced the rate of P application required to produce the maximum vegetative growth and N contents.

FIFTH SECTION: ROUND TABLE

26. Conclusions

Over four years have passed since the first international gathering of _Azolla_ researchers at IRRI as part of the Symposium on Nitrogen and Rice. Since that time interest in _Azolla_ research has increased dramatically. The large number of scientists conducting _Azolla_ research underscores the need for standardization of experimental results. Measurements of _Azolla_ biomass should include the washed dry weight of the _Azolla_ inoculum and incremental and/or the final biomass obtained. Total nitrogen content of _Azolla_ should be reported if possible. A simple and easily understood index of growth is given by the formula for relative growth rate (rgr):

$$rgr = \frac{Log_e\ W_f - Log\ W_i}{t}$$

where W_i and W_f are, respectively, the initial and final washed dry weights and t in the time elapsed (usually in days) between these two weight measurements. Results are usually expressed in $g.g.^{-1}.day^{-1}$ and will range from 0 to 0.350 for _Azolla_. Using rgr, biomass doubling time may be calculated by the formula:

$$t\ doubling = \frac{Log_e\ 2}{rgr}$$

Many constraints to the utilizations of _Azolla_ could be overcome if plant varieties with the following characteristics could be discovered or produced: tolerance to high and/or low temperature, desiccation, salinity, low phosphorus and high iron requirements and resistance to pests and disease. There is overlap between the distribution of _Azolla_ and regions characterized by one or more of these problems. Highest priority should be given to the collection, care, and screening of _Azolla_ from these regions. _Azolla_ with significant tolerance to one or more of these conditions should be dispersed to other laboratories for confirmation and germplasm maintenance. However, international transfer should be contemplated only after the _Azolla_ has been substantialy freed of contaminants. Plant pathology studies to determine if _Azolla_ may serve as a vector for disease should be considered an integral part of the screening process. Without a substantial effort to collect, screen, and maintain unique _Azolla_ populations the effort to understand the _Azolla_ breeding system will be of little practical value.

Significant manual labor is needed to cultivate _Azolla_. Mechanization is essential to practical application of _Azolla_ in both developing and developed countries and should be a strong second priority. Mechanization research should begin with development of small, efficient hand tools and progress to small scale motorized equipment to collect, disperse, and incorporate _Azolla_.

Specific _Azolla_ management schemes will vary depending upon the availability and cost of labor, water, inorganic nutrients (particularly nitrogen and phosphorus), propagation space, and rice production schedules. Development of these schemes should be a national or regional priority. Concise and conclusive Azolla research has resulted when an initial effort was made to determine the climatic and nutritional suitability of Azolla species and strains for a given region followed by assessment of the basic response of rice to Azolla nitrogen. An elementary cost analysis may be derived providing these data are used to design experiments which attempt to optimize the labor and nutrients necessary to produce available levels of Azolla nitrogen with an increased rice yield.

Most Azolla research, and the focus of this workshop, has been directed toward providing nitrogen for rice. Important potential applications also exist for wastewater treatment, weed control, and the production of high quality fodder and compost. Multiple use of Azolla will probably prove essential to the economic viability of local Azolla cultivation projects.

A respected international agency dealing with agriculture should be sought to endorse these proposals. The endorsed letter could then be sent to countries believed to be potential beneficiaries of Azolla cultivation.

S. N. Talley
Roundtable Chairman

List of participants

F. Abruña, USDA/ARS, Agricultural Experiment Station, College of Agricultural Sciences, Mayaguez Campus, Mayaguez, Puerto Rico 00708

S. Alemañy, Chancellor, University of Puerto Rico, Mayaguez Campus, Mayaguez, Puerto Rico 00708

L.A.T. Amador, Servicio de Extensión Agrícola, Centro Gubernamental, Edificio A. Altos, Arecibo, Puerto Rico 00612

F.L. Arens, FAO, Via delle Terme di Caracalla, Rome, Italy

A. Ayala, Dean and Director, College of Agricultural Sciences, University of Puerto Rico, Mayaguez, Puerto Rico 00708

A.J. Beale, Department of Agronomy and Soils, Agricultural Experiment Station, Venezuela Contract Station, Río Piedras, Puerto Rico 00931

H.E. Calvert, C.F. Kettering Res. Lab., 150 East South College St., Yellow Springs, Ohio 45387

R. Caudales, Department of Agronomy and Soils, University of Puerto Rico, Mayaguez, Puerto Rico 00708

R. Chevres-Román, Department of Agronomy and Soils, University of Puerto Rico, Mayaguez, Puerto Rico 00708

L.M. Cruz-Pérez, Director, Department of Agronomy and Soils, University of Puerto Rico, Mayaguez, Puerto Rico 0078

H.F. Diara, ADRAO/WARDA, B.F. 29 Richard Toll, Senegal, Africa

R. González-Trabal, Corporación Arrocera de Puerto Rico, AFDA, P. O. Box 9560, Cotto Station, Arecibo, Puerto Rico 00613

R. Guerrero, Department of Agronomy and Soils, University of Puerto Rico, Mayaguez, Puerto Rico 00708

M.I. Idoe, Rice Research and Breeding Station, P. O. Box 26, New Nickerie, Surinam

C. Kanareugsa, Rice Fertilization Research Branch, Rice Division, Dept. of Agriculture, Bangkhen, Bangkok 9, Thailand

A. Liogier, Jardín Botánico, Universidad de Puerto Rico, Venezuela Contract Station, Río Piedras, Puerto Rico 00931

J. López-Rosa, Director, Crop Protection Department, University of Puerto Rico, Mayaguez, Puerto Rico 00708

H. Lugo-Mercado, Department of Agronomy and Soils, University of Puerto Rico, Mayaguez, Puerto Rico 00708

T.A. Lumpkin, University of Hawaii at Manoa, Department of Agronomy Soil Science, 3190 Maile Way, Honolulu, Hawaii 96822

P. Meléndez, Associate Dean, School of Agricultural Sciences, University of Puerto Rico, Mayaguez, Puerto Rico 00708

M.A. Mercado, Corporación Arrocera, Central Cambalache, Arecibo, Puerto Rico 00612

A. Moretti, Instituto di Botanica, Facolta di Science, Via Foria 223 Napoli. Cod. Post 80139, Italy

O. Muñiz-Torres, Department of Agronomy and Soils, University of Puerto Rico, Mayaguez, Puerto Rico 00708

J.W. Newton, Northern Regional Research Center, USDA/ARS, 1815 North University, Peoria, Illinois 61604, USA

R.F. Olmeda, Urb. El Parque C-9, Barceloneta, Puerto Rico 00617

J.J. Peña-Cabriales, CIEA del IPN, Unidad Irapuato, Blvd. Gustavo Díaz Ordaz 237, Oficina 101, Apartado Postal 629, 36500, Irapuato, Gto., México

G.A. Peters, C.F. Kettering Res. Lab., 150 East South College Street, Yellow Springs, Ohio 45387, USA

L.A. Picó, Programa de Arroz, Secretaria de Agricultura, San Juan, Puerto Rico 00910

C.T. Ramírez, Centro Experimental de Lajas, Estación Experimental Agrícola, Recinto Universitario de Mayaguez, Apartado de Correos 940, Lajas, Puerto Rico 00667

K.R. Reddy, Agricultural Research and Education Center, Sanford, Institute of Food and Agricultural Sciences, Box 909, Sanford, Florida 32771, USA

P.A. Reynaud, ORSTOM, E.P. 1386, Dakar, Senegal

C. Rodríguez, Department of Marine Sciences, University of Puerto Rico, Mayaguez, Puerto Rico 00708

R.L. Rodríguez, Department of Crop Protection, University of Puerto Rico, Venezuela Contract Station, Río Piedras, Puerto Rico 00931

M.J.A. Sampaio, CENA, Av. Centenario S/N, Piracicaba, SP, CEP 13400, Brasil

J.I. Sánchez, Department of Agricultural Education, School of Agricultural Sciences, University of Puerto Rico, Mayaguez, Puerto Rico 00708

E.C. Schroder, Department of Agronomy and Soils, University of Puerto Rico, Mayaguez, Puerto Rico 00708

W.S. Silver, Department of Biology, University of South Florida, Tampa, Florida 33620

V. Snyder, Department of Agronomy and Soils, University of Puerto Rico, Mayaguez, Puerto Rico 00708

A. Sotomayor, Director, Mayaguez Institute of Tropical Agriculture, USDA/SEA, P. O. Box 70, Mayaguez, Puerto Rico 00709

C. Stearn, INTSOY, Department of Agronomy and Soils, University of Puerto Rico, Mayaguez, Puerto Rico 00708

S.N. Talley, 4347 Stollwood Drive, Carmichael, California 95608, USA

T.T. Trinh, AGPC Room C766, FAO, Via delle Terme di Caracalla, Rome 00100, Italy

C. Van Hove, University Catholique de Louvain, Lab. de Physiologie Vegetable, Place Croix de Sud. No. 4 1348 Louvain-la-Neuve, Belgium

A. Vélez-Ramos, Assistant Director, Estación Experimental Agrícola, University of Puerto Rico, Mayaguez, Puerto Rico 00708

Author Index

This index lists the names of the authors of the papers. The figures refer to the first pages of each paper.

Key words index

This index lists the key words given at the head of the papers. The figures refer to the first pages of papers, not to the pages on which the words are mentioned.